T0215784

Reconstruction-Free Compressive Vision for Surveillance Applications

Synthesis Lectures on Signal Processing

Synthesis Lectures in Signal Processing publishes 80- to 150-page books on topics of interest to signal processing engineers and researchers. The Lectures exploit in detail a focused topic. They can be at different levels of exposition—from a basic introductory tutorial to an advanced monograph—depending on the subject and the goals of the author. Over time, the Lectures will provide a comprehensive treatment of signal processing. Because of its format, the Lectures will also provide current coverage of signal processing, and existing Lectures will be updated by authors when justified.

Lectures in Signal Processing are open to all relevant areas in signal processing. They will cover theory and theoretical methods, algorithms, performance analysis, and applications. Some Lectures will provide a new look at a well established area or problem, while others will venture into a brand new topic in signal processing. By careful reviewing the manuscripts we will strive for quality both in the Lectures' contents and exposition.

Reconstruction-Free Compressive Vision for Surveillance Applications
Henry Braun, Pavan Turaga, Andreas Spanias, Sameeksha Katoch, Suren Jayasuriya, and Cihan Tepedelenlioglu
2019

Smartphone-Based Real-Time Digital Signal Processing, Second Edition
Nasser Kehtarnavaz, Shane Parris, and Abhishek Sehgal
2018

Anywhere-Anytime Signals and Systems Laboratory: From MATLAB to Smartphones, Second Edition
Nasser Kehtarnavaz, Fatemeh Saki, and Adrian Duran
2018

Anywhere-Anytime Signals and Systems Laboratory: From MATLAB to Smartphones
Nasser Kehtarnavaz and Fatemeh Saki
2016

Smartphone-Based Real-Time Digital Signal Processing
Nasser Kehtarnavaz, Shane Parris, and Abhishek Sehgal
2015

An Introduction to Kalman Filtering with MATLAB Examples
Narayan Kovvali, Mahesh Banavar, and Andreas Spanias
2013

Sequential Monte Carlo Methods for Nonlinear Discrete-Time Filtering
Marcelo G.S. Bruno
2013

Processing of Seismic Reflection Data Using MATLAB™
Wail A. Mousa and Abdullatif A. Al-Shuhail
2011

Fixed-Point Signal Processing
Wayne T. Padgett and David V. Anderson
2009

Advanced Radar Detection Schemes Under Mismatched Signal Models
Francesco Bandiera, Danilo Orlando, and Giuseppe Ricci
2009

DSP for MATLAB™ and LabVIEW™ IV: LMS Adaptive Filtering
Forester W. Isen
2009

DSP for MATLAB™ and LabVIEW™ III: Digital Filter Design
Forester W. Isen
2008

DSP for MATLAB™ and LabVIEW™ II: Discrete Frequency Transforms
Forester W. Isen
2008

DSP for MATLAB™ and LabVIEW™ I: Fundamentals of Discrete Signal Processing
Forester W. Isen
2008

The Theory of Linear Prediction
P. P. Vaidyanathan
2007

Nonlinear Source Separation
Luis B. Almeida
2006

Spectral Analysis of Signals: The Missing Data Case
Yanwei Wang, Jian Li, and Petre Stoica
2006

Reconstruction-Free Compressive Vision for Surveillance Applications

Henry Braun, Pavan Turaga, Andreas Spanias, Sameeksha Katoch, Suren Jayasuriya, and Cihan Tepedelenlioglu

ISBN: 978-3-031-01413-0   paperback
ISBN: 978-3-031-02541-9   ebook
ISBN: 978-3-031-00334-9   hardcover

DOI 10.1007/978-3-031-02541-9

A Publication in the Springer series
SYNTHESIS LECTURES ON SIGNAL PROCESSING

Lecture #17
Series ISSN
Print 1932-1236   Electronic 1932-1694

Reconstruction-Free Compressive Vision for Surveillance Applications

Henry Braun, Pavan Turaga, Andreas Spanias, Sameeksha Katoch, Suren Jayasuriya, and Cihan Tepedelenlioglu
Arizona State University

SYNTHESIS LECTURES ON SIGNAL PROCESSING #17

ABSTRACT

Compressed sensing (CS) allows signals and images to be reliably inferred from undersampled measurements. Exploiting CS allows the creation of new types of high-performance sensors including infrared cameras and magnetic resonance imaging systems. Advances in computer vision and deep learning have enabled new applications of automated systems. In this book, we introduce reconstruction-free compressive vision, where image processing and computer vision algorithms are embedded directly in the compressive domain, without the need for first reconstructing the measurements into images or video. Reconstruction of CS images is computationally expensive and adds to system complexity. Therefore, reconstruction-free compressive vision is an appealing alternative particularly for power-aware systems and bandwidth-limited applications that do not have on-board post-processing computational capabilities. Engineers must balance maintaining algorithm performance while minimizing both the number of measurements needed and the computational requirements of the algorithms. Our study explores the intersection of compressed sensing and computer vision, with the focus on applications in surveillance and autonomous navigation. Other applications are also discussed at the end and a comprehensive list of references including survey papers are given for further reading.

KEYWORDS

compressed sensing, sparse representations, track-before-detect, deep learning, surveillance

Contents

Preface

The book was inspired by a series of projects in the SenSIP Center and the School of Electrical, Computer and Energy Engineering at Arizona State University. The book is intended to be an introduction to the field of Compressive Vision. A comprehensive list of references is provided highlighting foundations and recent developments in this area. It covers several topics at the intersection of compressive sensing and computer vision with applications to surveillance. Particular attention is given to detection and tracking as well as classification problems.

Henry Braun, Pavan Turaga, Andreas Spanias, Sameeksha Katoch, Suren Jayasuriya, and Cihan Tepedelenlioglu
March 2019

Acknowledgments

The authors have been supported in part from the SenSIP center and the NSF I/UCRC grant 1540040. Big thanks to Alphonso Samuel of Raytheon Missile Systems who reviewed parts of this work.

Henry Braun, Pavan Turaga, Andreas Spanias, Sameeksha Katoch, Suren Jayasuriya, and Cihan Tepedelenlioglu
March 2019

CHAPTER 1

Introduction

In this book, we present work at the intersection of compressed sensing (CS), image processing, and computer vision. Using the single-pixel camera [1] as a model, we discuss inference algorithms for CS data. Two broad categories of inference problems are considered: video-based target tracking, and image-based detection/classification [2]. In particular, we focus on the field of **reconstruction-free compressive vision** where image processing and computer vision algorithms are developed directly in the compressed domain itself, rather than on reconstructed images.

CS is a relatively recent paradigm in sensor and signal processing which combines specially-designed sensors with *a priori* knowledge of a signal's compressibility. By exploiting this knowledge, a signal of interest can be recovered from an extremely underdetermined set of measurements [3]. The potential benefits are clear: compressive sensing technology results in lower data rate requirements and less costly sensing hardware. However, these advantages come at the cost of higher computational requirements, as measurements from CS sensors must be processed to reconstruct the signal of interest.

An important example of a CS sensor is the single-pixel camera [1, 4]. The single-pixel camera, in its simplest embodiment, displays a pseudo-random measurement pattern on a digital micromirror device. The image to be sensed is projected on this micromirror and the reflected light is captured by a photodiode, resulting in an analog implementation of the inner, or "dot," product operation. By rapidly switching through a known set of measurement patterns, a vector of these inner products with the image is assembled. These sensor measurements do not look like natural images, but they contain a rich coded representation of the visual scene that can be exploited in further computational post-processing.

From a set of CS measurements, a reconstruction algorithm is typically used to estimate the image. Natural images are known to be compressible—see for instance coding methods the discrete cosine or wavelet transforms [5]. The transformed images contain most of the signal energy in a very small fraction of the components, with most of the components having near zero magnitude. The CS reconstruction procedure identifies and estimates these large components, allowing the image to be recovered from the single-pixel's camera's measurements. Several variations on this theme exist, but all rely on exploiting prior knowledge of a signal's *sparsity* in a transformed domain. The number of measurements required to accurately reconstruct the signal is larger than the number of nonzero components of the transformed signal; additional measurements are required to identify which indices are non-zero. The total number of measurements, though, is far less than the total length of the signal to be reconstructed. We provide an overview of compressed sensing and recovery algorithms in Chapter 2.

1.1   TARGETED APPLICATIONS

This work targets automated systems which may benefit from the use of CS imaging. Surveillance is one such application: for example, by adopting computationally simple object recognition algorithms using CS imaging, the cost of IR video monitoring can be reduced. This reduction comes partially through lower computational demands, but also allows data transmission to be dramatically reduced. If the relevance of data can be determined at the sensor, only relevant data must be transmitted to a central location.

Another application is autonomous vehicles, especially unmanned aerial vehicles (UAVs). More effective classification algorithms for UAV applications allow higher performance or lower sensor cost. By reducing the computational, and therefore monetary, cost of a UAV's situational awareness, new applications for autonomous UAVs become feasible.

1.2   PROBLEM MOTIVATION

Improving video analytics is especially important for applications in the surveillance domain. However, CS reconstruction of video sequences is less straightforward than for still images. Individual frames are compressible in a linear basis, but statistical dependence between frames is not as easily described. Even a relatively simple model of inter-frame image motion is bilinear rather than linear in the image intensity estimated motion vectors [6]. Non-CS video compression algorithms divide video sequences into Intra-coded *I frames* and predicted *P frames*, but extending this approach to CS is non-trivial since we have no access to the images from which predicted motion is estimated. Exploiting the dependence between frames improves video reconstruction performance, but finding the best possible method remains an active area of research.

As described above, recovery of signals from CS measurements is computationally demanding. However, in some cases, such as automated surveillance or navigation, a system may not require a human to observe a video at all. In these cases it may be possible to perform target recognition and other computer vision tasks directly on the compressively sensed data, without reconstructing video frames. This is what we name **reconstruction-free compressive vision** and it is a challenging problem since the multiplexing effect of the compressed sensing operation forces the researcher to abandon principles, such as shift invariance, on which conventional algorithms rely heavily.

Consider the problem of tracking an object in a video sequence. This is already a difficult computer vision task without the addition of CS. Targets change scale as they get closer and further from the sensor. Light conditions change as they move in and out of shadows; reflections and glare cause sudden changes in the perceived color of the target's surface. Non-rigid objects change not just their projection into the image plane, but their 3D shape as well. None of these effects can be fully predicted from a single image of the target.

CS-based tracking adds to these difficulties, leaving a smaller set of tools. In-plane translation and rotation, normally relatively easily modeled, become much more difficult. Given a CS measurement vector of an image containing an object, we cannot even predict how the measurements would change if the object moves one pixel to the left. Doing so would require reconstructing the image, defeating our purpose of speeding up computation by avoiding reconstruction. A region of interest cannot be easily cropped out of an image either; every CS measurement encodes information about the whole image. On the other hand, some operations are still available for use. Vector lengths, Euclidean distances, and inner products are all preserved under the compressed sensing operation, albeit with increased noise. A reconstruction-free computer vision algorithm must use this reduced set of tools.

It is the goal of this book to outline a vision and state-of-the-art work in the area of reconstruction-free computer vision algorithms. We focus on two main types of algorithms: detection/tracking as well as image classification. We introduce a comprehensive survey of different research directions including new state-of-the-art deep learning methods applied to CS measurements to perform direct inference. By leveraging modern computational tools, reconstruction-free compressive vision can offer a low-cost appealing alternative to computationally expensive and energy-hungry traditional methods.

1.3   ORGANIZATION

The book is organized as follows. First, Chapter 2 presents an overview of CS and its mathematical formulation. Hardware implementations of CS imaging sensors are also described, including the single-pixel camera considered in this work. An overview of several CS reconstruction algorithms is given, and their performance is compared. Chapter 3 discusses existing work in image and video processing and computer vision focused on surveillance applications. Several computer vision algorithms for detection and classification of images are discussed. The problem of visual object tracking is described and some algorithms for tracking and state estimation are presented. Chapter 4 discusses the work lying at the intersection of compressed sensing and computer vision. It elaborates on reconstruction-free compressive tracking algorithm that uses the track-before-detect paradigm. It also shows recent work on direct inference on compressed measurements for image classification. Chapter 5 summarizes the research directions established in the booklet and discusses the possible future avenues that can be explored.

CHAPTER 2

Compressed Sensing Fundamentals

The work presented in this book lies at the intersection of CS and computer vision for surveillance applications. We attempt to separate these two areas by devoting this chapter to CS theory and literature and Chapter 3 to related work in image processing and computer vision that we utilize for reconstruction-free compressive vision. We begin by introducing the CS sensing matrix, ℓ_1 reconstruction, and other fundamental background information in Section 2.1. Next, Section 2.3 discusses existing CS imaging hardware in order to provide real-life motivation for the work. CS reconstruction algorithms are then discussed in Section 2.4 and bounds on CS sensing performance are covered in Section 2.6. Finally, deep learning for CS reconstruction is presented in Section 2.8.

2.1 FOUNDATIONS OF COMPRESSED SENSING

In this section, the background theory of compressive sensing is described. Terminology and symbols used throughout the work are introduced and their significance is explained. In essence, compressive sensing is the use of prior knowledge of a signal's statistics to reduce the number of measurements required to recover the signal. The use of this prior knowledge allows the Nyquist limit to be effectively "broken." CS reconstruction is concerned with the recovery of a compressible signal of interest $\mathbf{x} \in \mathbb{R}^{N \times 1}$ from a related measurement vector $\mathbf{y} \in \mathbb{R}^{M \times 1}$. $\mathbf{x}$ and $\mathbf{y}$ are related by a linear transformation,

$$\mathbf{y} = \Phi\mathbf{x} + \mathbf{w}, \tag{2.1}$$

where $\mathbf{w} \in \mathbb{R}^{M \times 1}$ is a noise vector and $\Phi \in \mathbb{R}^{M \times N}$ is known as the sensing matrix.

If $M < N$ (i.e., $\mathbf{y}$ is shorter than $\mathbf{x}$), the system in (2.1) is under-determined and x cannot be recovered with certainty from y even in the noiseless ($\mathbf{w} = \mathbf{0}$) case. However, let $\mathbf{x}$ be k-sparse (at most k non-zero components) with respect to some known basis with inverse transform matrix Ψ. That is,

$$\mathbf{y} = \Phi\mathbf{x} + \mathbf{w} = \Phi\Psi\boldsymbol{\theta} + \mathbf{w}, \tag{2.2}$$

$$\|\boldsymbol{\theta}\|_0 \leq k, \tag{2.3}$$

where $\boldsymbol{\theta}$ is the transformed representation of $\mathbf{x}$. Here the ℓ_0 norm denotes the number of non-zero elements. Strictly speaking, this is not a norm as it does not satisfy the homogeneity property. However, this mild abuse of notation is common in the literature and will be used in the remainder of this document. The basis matrix Ψ should have full row rank ($\mathrm{rank}(\Phi) = N$), but need not be square. Overcomplete bases are commonly used in several applications [7, 8]. Let $\mathrm{A} = \Phi\Psi$ be the combined sensing operator; this notation is used to simplify equations in which the two appear together.

If the conditions in (2.2) and (2.3) are met, CS theory dictates that it can be recovered with high probability as long as

$$(1 - \delta_k)\|\boldsymbol{\theta}\|_2 \leq \|\mathrm{A}\boldsymbol{\theta}\|_2 \leq (1 + \delta_k)\|\boldsymbol{\theta}\|_2 \tag{2.4}$$

for all k-sparse $\mathbf{x}$, with sufficiently small δ_k. (2.4) is known as the restricted isometry property (RIP) [9].

Several pseudorandom sensing matrices satisfy the RIP with high probability, including an i.i.d. Gaussian matrix [10]. Other matrices which satisfy the RIP include the Fourier and Hadamard ensemble [11] and the random orthoprojector [12]. The random orthoprojector is noteworthy as finding its pseudoinverse is computationally trivial ($\Phi^\dagger = \Phi^T (\Phi\Phi^T)^{-1} = \Phi^T$), a desirable property for some algorithms. In [13] an attempt is made to develop a deterministic sensing matrix, but the given approach does not fully satisfy the RIP.

Few naturally occurring signals are perfectly sparse, but many are known to be compressible; we here focus on natural images and video sequences. The JPEG image compression algorithm relies on the observation that the discrete cosine transform (DCT) of a natural image results in a transform-domain signal whose coefficients follow a power law distribution, with signal energy concentrated in a small number of components and the remainder near zero [14]. The JPEG2000 standard similarly relies on the discrete wavelet transform for compression [15]. In [16], discussed in Section 2.2.2, a bound is given for reconstruction of signals which are not perfectly sparse.

If the RIP is satisfied, $\boldsymbol{\theta}$, and therefore $\mathbf{x}$, can be recovered in the noiseless case from y by solving the ℓ_0 optimization problem given by

$$\begin{aligned} \underset{\boldsymbol{\theta}}{\text{minimize}} \quad & \|\boldsymbol{\theta}\|_o \\ \text{subject to} \quad & \mathbf{y} - \mathrm{A}\boldsymbol{\theta} = 0. \end{aligned} \tag{2.5}$$

Unfortunately this optimization is nonconvex and cannot be solved in polynomial time. However, the convex relaxation of this problem is much more easily solved. It is given by

$$\begin{aligned} \underset{\boldsymbol{\theta}}{\text{minimize}} \quad & \|\boldsymbol{\theta}\|_1 \\ \text{subject to} \quad & \mathbf{y} - \mathrm{A}\boldsymbol{\theta} = 0. \end{aligned} \tag{2.6}$$

where the ℓ_0 norm is replaced with the ℓ_1 norm. This problem is known as basis pursuit (BP). It is convex and can be solved with conventional convex optimization techniques.

2.2 RELATED CONVEX PROBLEMS

The basis pursuit reconstruction problem described in Section 2.1 is only the simplest of many related problems. In this section extensions and related problems are described. Several other convex optimizations exist; a selection of these are covered. Model-based CS, discussed in Section 2.2.2, improves performance by taking advantage of additional information not captured by a sparse prior. Blind CS, discussed in Section 2.2.3, treats the case where the sensing matrix is unknown or incompletely known.

Basis pursuit is valid in the noiseless case when $\mathbf{y} = A\boldsymbol{\theta}$ must be satisfied exactly. Other related convex algorithms for noisy data include basis pursuit denoising (BPDN) and lasso. BPDN relaxes the equality constraint in (2.6), replacing it with an ℓ_2 term in the optimization which penalizes deviation from $\mathbf{y} = \Phi\Psi\boldsymbol{\theta}$. The BPDN problem is given by

$$\underset{\boldsymbol{\theta}}{\text{minimize}} \, \|\mathbf{y} - A\boldsymbol{\theta}\|_2 + \tau \, \|\boldsymbol{\theta}\|_1 \,, \tag{2.7}$$

where τ is a tuning parameter reflecting the trade-off between encouraging sparsity and promoting a close match between the estimate and measured data. Intuitively, if measurements are noisy or there is a strong prior knowledge of sparsity of $\boldsymbol{\theta}$, a larger τ will be chosen. Conversely, if measurement noise is small or $\boldsymbol{\theta}$ is not known to be highly sparse, a smaller τ will be chosen. In additive white Gaussian noise (AWGN) with known signal and noise power, τ may be optimized as in (2.17), discussed in Section 2.2.1. In practice, however, the solution is not highly sensitive to the choice of τ.

In the lasso algorithm, the objective function and constraint are reversed relative to BP. Where the BP algorithm seeks the sparsest possible $\boldsymbol{\theta}$ given the data, lasso seeks the closest possible match with the data subject to a sparsity constraint. Lasso solves the optimization problem,

$$\begin{aligned} \underset{\boldsymbol{\theta}}{\text{minimize}} \quad & \|\mathbf{y} - A\boldsymbol{\theta}\|_2 \\ \text{subject to} \quad & \|\boldsymbol{\theta}\|_1 \le \epsilon_1, \end{aligned} \tag{2.8}$$

where ϵ_1 is a tuning parameter controlling the trade-off between sparsity of $\boldsymbol{\theta}$ and a close match to measurements $\mathbf{y}$.

With the appropriate choice of tuning parameters, all three problems become equivalent. In the limit of (2.7) as $\tau \to 0$, the solution of (2.6) is recovered and BPDN and BP are equivalent. Likewise, for every τ in (2.7), some ϵ_1 exists which results in the same solution to (2.8).

2.2.1 CS RECONSTRUCTION AS BAYESIAN INFERENCE

In Section 2.1, convex reconstruction based on the ℓ_1 norm was presented as a relaxation of a non-convex ℓ_0 problem. An alternative view of the problem [17] as maximum *a posteriori* (MAP) estimation under a Laplace prior is described here. This formulation will be useful in discussion of graphical reconstruction methods in Section 2.4.2. Let $\boldsymbol{\theta}$ be distributed i.i.d. Laplace with zero mean and scale parameter b. Its PDF is given as,

$$p(\boldsymbol{\theta}) = \prod_{n=1}^{N} \frac{1}{2b} \exp\left(-\frac{|\theta_n|}{2b}\right) = \left(\frac{1}{2b}\right)^N \exp\left(-\frac{\|\boldsymbol{\theta}\|_1}{2b}\right). \qquad (2.9)$$

The choice of a Laplace distribution for $p(\boldsymbol{\theta})$ is motivated by its history as a sparseness-inducing prior [18]; it is clearly a more informative prior than the Gaussian used in least squares.

$\mathbf{y}$ is then modeled as in (2.2) as a multiplication with the sensing matrix plus additive i.i.d. Gaussian noise of variance σ_w^2. The conditional PDF of $\mathbf{y}$ is given as

$$p(\mathbf{y}|\boldsymbol{\theta}) = \frac{1}{\left(2\pi\sigma_w^2\right)^{\frac{M}{2}}} \exp\left(-\frac{1}{2\sigma_w^2} \|\mathbf{y} - \boldsymbol{\Phi\Psi\theta}\|_2\right). \qquad (2.10)$$

The posterior PDF is then derived using Bayes' rule as follows:

$$p(\boldsymbol{\theta}|\mathbf{y}) = \frac{p(\mathbf{y}|\boldsymbol{\theta})p(\boldsymbol{\theta})}{p(\mathbf{y})} \qquad (2.11)$$

$$= \frac{1}{p(\mathbf{y})} \left(\frac{1}{2b}\right)^N \frac{1}{\left(2\pi\sigma_w^2\right)^{\frac{M}{2}}} \exp\left(-\frac{1}{2\sigma_w^2} \|\mathbf{y} - \boldsymbol{\Phi\Psi\theta}\|_2 - \frac{\|\boldsymbol{\theta}\|_1}{2b}\right). \qquad (2.12)$$

From this posterior PDF we derive an expression for the MAP estimate,

$$\hat{\boldsymbol{\theta}} = \arg\max_{\boldsymbol{\theta}} p(\boldsymbol{\theta}|\mathbf{y}) \qquad (2.13)$$

$$= \arg\max_{\boldsymbol{\theta}} \log\left(p(\boldsymbol{\theta}|\mathbf{y})\right) \qquad (2.14)$$

$$= \arg\min_{\boldsymbol{\theta}} \left(\frac{1}{2\sigma_w^2} \|\mathbf{y} - \boldsymbol{\Phi\Psi\theta}\|_2 + \frac{\|\boldsymbol{\theta}\|_1}{2b}\right) \qquad (2.15)$$

$$= \arg\min_{\boldsymbol{\theta}} \left(\|\mathbf{y} - \boldsymbol{\Phi\Psi\theta}\|_2 + \frac{\sigma_w^2}{b} \|\boldsymbol{\theta}\|_1\right). \qquad (2.16)$$

Setting τ as

$$\tau = \frac{\sigma_w^2}{b}, \qquad (2.17)$$

this is equal to the BPDN optimization problem of (2.7). This Bayesian interpretation of CS is heavily used in discussion of message passing algorithms, discussed in Section 2.4.2.

2.2.2   MODEL-BASED COMPRESSED SENSING

In model-based CS, the constraint on number of non-zero components $k = \|\boldsymbol{\theta}\|_0$ is replaced by a more sophisticated model [16]. The most notable models are block-sparse and tree-sparse signals. A bound on the performance of ℓ_1 BP in reconstruction of approximately sparse data, distributed according to a power law, is also given.

The traditional CS approach constrains the set of possible signals, but still allows the signal θ to lie on any of the $\binom{N}{k}$ possible k-sparse subspaces, with equal likelihood. For natural signals where this is too broad, model-based CS further restricts the allowed subspaces, allowing reconstruction from fewer measurements (i.e., smaller N). The RIP is replaced with a model-based RIP which defines isometry only for the allowed subspaces.

2.2.3   BLIND COMPRESSED SENSING

Blind CS [19] considers the case where both the (transformed) signal of interest θ and sparsity basis Ψ are unknown. At first glance, this may seem an impossible task, and in fact in general it is. In spite of this, the sparse prior on θ can be combined with additional constraints on Ψ to allow recovery of the signal x from an underdetermined set of measurements. This requires collecting multiple signals which are sparse under the same sparsity basis Ψ and jointly determining the basis for all of them. While blind CS achieves correct reconstructions with high probability, it requires higher sensing rates than the non-blind case.

Three constraints on the basis are considered in [19]. In the first, Ψ is known to be one of a finite and known set of commonly occurring bases, for instance one of several possible wavelet transforms. This case can be treated as a series of CS problems. In the second case, the rows of the basis are known to be a subset of a much larger dictionary matrix. This problem can be solved using standard CS methods, treating the larger dictionary as an overcomplete sparse basis. In the final case, Ψ is block-diagonal and the problem is solved by reformulating it as a dictionary learning problem.

2.3   SENSORS FOR COMPRESSIVE IMAGE AND VIDEO CAPTURE

This section describes several hardware implementations for capturing visual data using compressive sensing as the main sampling mechanism. The first major work is the single-pixel camera from Rice University, which helped inspire several future compressive sensing hardware to capture MRI, radar, hyperspectral data, and 4D light fields. We give a brief overview of these hardware platforms as they provide the motivation to do reconstruction-free compressive vision, especially in imaging regimes where sensor arrays are expensive to obtain high spatial resolutions.

2.3.1   SINGLE-PIXEL CAMERA

Researchers at Rice University have developed a prototype single-pixel camera [1, 20] and complementary algorithms for processing of its output. The camera replaces the focal plane array of a conventional camera with a digital micromirror device (DMD) and a single light sensor. The DMD displays a series of pseudorandom masks corresponding to the rows of the measurement matrix Φ, effectively performing a series of inner products on the incoming image. The modulated light is then focused on a photodiode, which measures the magnitude of this inner product. By taking several measurements of this single pixel in time, then various reconstruction algorithms for compressive sensing can be performed to recover the original image. The advantages of a single-pixel camera are primarily suited for imaging domains where multiple pixels is prohibitive or expensive due to fabrication costs, such as terahertz or far-infrared spectrum, and thus spatial resolution can be recovered in post-processing. Some disadvantages of the single-pixel camera include the need for multiple measurements in time which require static scenes, and the processing time needed to recover the image. The single-pixel camera concept described above is being commercialized by InView corporation. InView's short-wave infrared (SWIR) cameras are capable of operating in a high-resolution/low-framerate and low-resolution/high-framerate mode. InView's products are not truly single-pixel: multiple measurements are sensed in parallel, reducing the time to acquire a single image [21].

2.3.2   INFRARED AND HYPERSPECTRAL COMPRESSIVE SENSING CAMERAS

Researchers have exploited the need for less pixels/detectors in compressive sensing to extend into infrared and hyperspectral imaging, while reducing the cost of the hardware platforms. FPA-CS [22] introduced an array of 64×64 parallel single-pixel detectors sensitive in the short-wave infrared (SWIR), and utilized a DMD to compressively measure the scene at each of these detectors. The resulting reconstruction was able to achieve 1 megapixel resolution at video rates, capturing live SWIR images. The entire system cost under \$4,000, which is considerably cheaper than a \$60,000 megapixel SWIR sensor.

Hyperspectral imaging has been of importance for many scientific and biomedical applications. For compressively sensing this 3D hypercube of spatial and spectral information, The CASSI system [23] employs coded aperture methods to multiplex spatial and spectral measurements onto the sensor. This system was able to mitigate existing tradeoffs between spatial and spectral resolution and light acquisition. This system was later extended to capture hyperspectral video at 30 frames per second [24].

2.3.3   MEDICAL IMAGING AND MRI

Some of the most valuable contributions of CS are in the field of medical imaging, and especially in magnetic resonance imaging (MRI) [25, 26]. Other applications include computed tomography (CT), positron emission tomography (PET), and Ultrasound imaging. The application of CS theory to MRI is briefly described in this section.

MRI exploits magnetic resonance, a process by which atoms in a magnetic field absorb and re-emit RF energy at a specific resonance frequency. In a typical MRI scanner, a very strong (typically 1–10 T) magnetic field is established in the scanner with a superconducting electromagnetic coil. Additional coils establish gradients in magnetic field strength over the field of view. A sequence of RF pulses is then transmitted into the field of view and the response of the object to be imaged is recorded [27, 28].

MRI is naturally amenable to CS techniques; the time and monetary cost per sample is high, creating an immediate need for techniques which reduce the number of samples needed. Sensing is also performed in k-space (i.e., Fourier domain) so reconstruction must be done whether or not CS is used. A conventional 3D MRI performs Cartesian sampling and outputs a "data cube" of samples evenly spaced on the three axes of k-space. A fast Fourier transform (FFT) is then needed to recover the 3D image for viewing. In parallel imaging implementations, more sophisticated reconstruction techniques are required to process the data from multiple receiver coils (sensors).

Designing the sensing matrix of an MRI is non-trivial. Sampling is limited by the maximum slew rate of the equipment used; successive samples must be near each other in k-space [29, 30]. In [31], the Cartesian data cube is simply undersampled. In [32–35], a radial trajectory was used, and in [36, 37] a spiral trajectory was used. [38] advocates a pseudorandom poisson-disc sampling approach; this method generates good incoherence between the sparse basis and the samples, while maintaining relatively even coverage of k-space.

With the imaging speedup available from CS, applications such as 4D MRI, which adds video to the 3D MRI image, become more feasible. In [39], a radial sensing trajectory was used to reconstruct a 4D MRI of the lungs during forced expiration. Acquisition time for each 3D frame was under 150 ms. "Bookend" image acquisition phases at the beginning and end of the expiration were used in which the patient is instructed to hold his or her breath while additional measurements are taken. The reconstruction algorithm minimizes spatial and temporal total variation and allows each frame to be expressed as a sum of the two bookend frames. This allows the full 4D signal to be reconstructed with only three measurements per 2D MRI "slice."

2.3.4   4D LIGHT FIELDS, DIGITAL HOLOGRAPHY, AND MICROSCOPY

Beyond traditional imaging scenarios mentioned above, compressive sensing has found success in exotic imaging domains where the visual signal of interest is difficult to extract, typically of higher dimensionality than the inherent sensing element, and is unable to be sensed natively at its Nyquist resolution. We briefly discuss these cases as futuristic use cases for compressive vision as these optical technologies develop and mature.

4D light fields represent the spatio-angular radiance distribution captured by a camera, and are utilized in computational imaging and graphics for depth mapping, digital refocusing, and novel view synthesis. However, capturing 4D light fields at high-spatial and angular resolution is a key challenge, as most sensors must multiplex angular information onto the spatial sensor using microlenses or other optical elements. This is a problem that is ideally suited for compressive sensing hardware platforms. Compressive light field imagers that utilize coded aperture masks [40] or diffractive gratings [41] have been shown to overcome these spatio-angular tradeoffs, achieving high-quality 4D light field capture.

Holography captures the full light field emanating from a scene into a format which can be later reproduced as a volumetric 3D display without the need for glasses. This is analogous to a set of microphones that record the sound of a room and then play it back in surround sound to the listener. Researchers recently have used results from compressive sensing to overcome the undetermined measurements in digital holography [42], and have extended it to perform compressive holographic video [43].

Finally, the field of microscopy is vitally important to biomedical imaging, and several tradeoffs between spatial resolution, depth-of-field, and light acquisition in the optics of microscopes exist. To overcome these limits, researchers have turned to encoding the lighting in a microscope to perform compressive sensing in the illumination of Fourier Ptychography [44]. As more emerging imaging domains are bottlenecked by tradeoffs between sampling resolution and signal-to-noise ratio and other optical metrics, compressive sensing will play an even more vital role in the future.

2.4   NON-CONVEX RECONSTRUCTION ALGORITHMS

Many solvers have been developed for compressive reconstruction. We here discuss greedy solvers and belief propagation approaches to the problem. Convex algorithms attempt to solve (2.6) or a related problem more quickly than general interior-point methods. These algorithms include GPSR [45] and SPGL1 [46]. Although they are many times faster than general convex solvers, these algorithms typically are more computationally expensive than greedy and belief propagation-based approaches, which we thus discuss in the following sections.

2.4.1   GREEDY SOLVERS

The greedy algorithms discussed here follow a procedure in which the signal support is iteratively built up. On each iteration, the elements of $\boldsymbol{\theta}$ which most improve the reconstruction are added to the support. Optionally, the least helpful components may also be dropped from the signal support. Three greedy algorithms are described here, although more exist.

Orthogonal Matching Pursuit

Orthogonal Matching Pursuit (OMP) and its related family of algorithms [47–50] approximately solve (2.5) in a greedy manner. OMP runs in $O(MNk)$ time and is in practice much faster than convex solvers. This speedup is achieved using an iterative greedy procedure in which the support of the estimated solution $\hat{\boldsymbol{\theta}}$ is steadily increased until a k-sparse $\hat{\boldsymbol{\theta}}$ is achieved. At each iteration, the residual $\mathbf{r} = \mathbf{y} - A\hat{\boldsymbol{\theta}}$ is calculated. The column $A_{:,j}$ of A which maximizes the inner product $\langle A_{:,j}, \mathbf{r} \rangle$ is added to the support of $\hat{\boldsymbol{\theta}}$. $\hat{\boldsymbol{\theta}}$ is then recalculated over the newly increased support. When $\|\boldsymbol{\theta}\|_0 = k$, the algorithm is halted and the current $\hat{\boldsymbol{\theta}}$ is output as the solution.

A theoretical guarantee exists for reconstruction by OMP, although it is weakened relative to the convex solvers. Specifically, for $\delta \in (0, 0.36)$ and $M \geq 4kln(N/\delta)$, OMP will identify the correct $\boldsymbol{\theta}$ with probability greater than $1 - 2\delta$ [48]. Note that this result is for the noiseless case ($\mathbf{w} = 0$).

CoSAMP

The Compressive Sampling Matched Pursuit (CoSAMP) reconstruction algorithm [51] is similar to OMP but adds the ability to discard unneeded elements from the support of $\hat{\boldsymbol{\theta}}$. On each iteration, some number of components are added to the support and least squares estimation is performed. The smallest components of the least-squares estimate are then discarded to maintain the desired sparsity level $\left\Vert \hat{\boldsymbol{\theta}} \right\Vert_0 = k$.

AdMIRA

The atomic decomposition for minimum rank approximation (ADMiRA) algorithm [52] applies an approach generalizing CoSAMP to the problem of reconstructing low-rank matrices rather than sparse vectors. This is useful for processing of video with lighting changes but no motion, since if each frame is assigned to a single column of a matrix, the resulting matrix will be low-rank.

2.4.2   GRAPHICAL METHODS

The compressed sensing architecture can be represented as a Bayesian Network. $\boldsymbol{\theta}$ is drawn from a sparse prior distribution, and $\mathbf{x}$ and $\mathbf{y}$ are then calculated based on $\boldsymbol{\theta}$. The problem can thus be represented as a cyclic causal model and solved accordingly. Generalized approximate message passing (GAMP) [53] is an algorithm based on a quadratic approximation of a loopy belief propagation (BP) algorithm. In loopy BP, the measured vector $\mathbf{y}$ passes messages to the elements of $\mathbf{x}$ consisting of the conditional probability of each possible value. Similarly, the sparse prior passes messages down to $\boldsymbol{\theta}$. GAMP may be viewed as a version of this message-passing algorithm in which only the means and variances of the messages are passed. In this section, the concept of Bayesian networks is first introduced. The belief propagation (BP) algorithm is described, followed by a review of its many extensions, generalizations, and approximations.

Bayesian Networks

A Bayesian network, or Bayes net, consists of a directed acyclic graph (DAG) in which nodes represent random variables and edges represent dependencies between nodes. Any node is conditionally independent from its non-neighboring nodes given its upstream and downstream neighbors. Although no directed cycles are allowed, *undirected* loops may be present. This constraint defines a clear path from causes (e.g., differing hypotheses) to effects (e.g., experimental measurements). Figure 2.1 shows a simple Bayesian network. In this example X_5 is conditionally independent from all other nodes given X_3. The goal of inference is to determine $p(X_i|E)$, the distribution of each unknown variable X_i given a set of known evidence variables E. Evidence

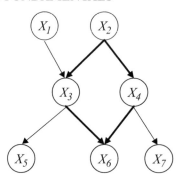

Figure 2.1: A simple Bayesian network.

may appear at any node in a network, but most typically is measured at the furthest downstream leaf nodes. Note also that the graph in Figure 2.1 has an undirected loop, shown in bold.

Performing inference on a Bayesian Network consists of estimating the marginal conditional probability distribution of a node or nodes (the unknowns), given values of a different set of nodes (the evidence). Performing exact inference on Bayesian networks has been shown to be in general NP-hard [54]. However, constrained versions of the problem have been solved exactly and good approximate solutions exist for many other cases. Belief propagation (BP),[1] discussed below, generates exact solutions for tree and polytree graphs (that is, graphs with no undirected loops), and leads to good approximations for many other problems. Further modifications of BP offer advantages in computational burden or stability.

Belief Propagation for Trees

Pearl's belief propagation algorithm [55] is a method for performing inference on Bayesian networks, Markov random fields, and factor graphs. In this section its operation on polytree-structured Bayesian networks is described. In graphs which are trees or polytrees, BP produces exact results. However, no such guarantee exists for graphs containing undirected loops.

Consider the fragment of a Bayesian network shown in Figure 2.2. This network consists of a node X whose value is to be estimated based on a set of downstream ("child") nodes $\{Y_1, Y_2, \ldots\}$ and upstream ("parent") nodes $\{U_1, U_2, \ldots\}$. The conditional probability distributions are known, and the upstream and downstream nodes are conditionally independent given X. The PDF of X can then be written in terms of the PDFs of its upstream and downstream neighbors as

$$p(x|u_1, u_2, \ldots, y_1, y_2, \ldots) = \frac{1}{\mathcal{Z}} \prod_i p(y_i|x) \int_{u_1, u_2, \ldots} p(x|u_1, u_2, \ldots) \prod_j p(u_j)\, du_1 du_2 \ldots,$$

(2.18)

---

[1]Note that basis pursuit and belief propagation share the acronym BP. Where ambiguity exists, the full name will be used to avoid confusion.

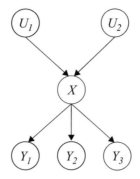

Figure 2.2: A fragment of a Bayesian network.

where $\mathcal{Z}$ is a normalizing constant.

In BP, each node passes messages to its upstream and downstream neighbors. Each message is a measure but is not a probability distribution; it does not in general integrate to one. In the upstream direction, each node Y_i sends a message $\lambda_{Y_i X}(x)$ to node X. Likewise, each upstream node U_i sends a message $\pi_{U_i X}(x)$. These λ and π messages take the place of the conditional probabilities in (2.18) to form a new expression for our belief about $p_X(x)$. Specifically,

$$BEL_X(x) = p(x|u_1, u_2, \ldots, y_1, y_2, \ldots) = \frac{1}{\mathcal{Z}}\lambda_X(x)\pi_X(x) \tag{2.19}$$

$$\lambda_X(x) = \prod_i \lambda_{Y_i X}(x) \tag{2.20}$$

$$\pi_X(x) = \int_{u_1, u_2, \ldots} p(x|u_1, u_2, \ldots) \prod_j \pi_{U_j X}(x)\, du_1 du_2 \ldots \tag{2.21}$$

In contrast to (2.18), $\pi_{U_j X}(x)$ and $\lambda_{Y_i X}(x)$ are not known exactly, but are based on the messages received at U_j and Y_i, respectively. Note that $\lambda_{Y_i X}(x)$ denotes a measure passed as an upstream message from child to parent, while $\lambda_X(x)$ refers to a measure calculated at node X from received λ messages. Similarly, $\pi_X(x)$ is calculated from received $\pi_{U_i X}(x)$ messages. The subscript X will be dropped for brevity in the following discussion unless it is needed to avoid ambiguity. Normalization of beliefs is not necessary but may improve numerical stability and prevent underflow problems.

In a tree or polytree network, the assumption of conditional independence above is satisfied and (2.18) holds. Inference then proceeds as follows. Evidence nodes (i.e., nodes whose values have been measured) are initialized. That is, for each evidence node E_i with known value

e_i,

$$\lambda_{E_i}(x) = \delta(x - e_i) \tag{2.22}$$
$$\pi_{E_i} = \delta(x - e_i) \tag{2.23}$$
$$BEL_{E_i}(x) = \delta(x - e_i). \tag{2.24}$$

Non-evidence nodes X_i with no parents are initialized using their Bayesian prior, as follows:

$$\pi_{X_i}(x) = p_{X_i}(x). \tag{2.25}$$

The downstream evidence $\lambda(x)$ of nodes with no children is initialized to 1. All other messages are unavailable at the start of the algorithm. Once initialized, belief updates proceed by iterating the following steps until convergence.

- If all upstream messages $\lambda_{Y_i X}(x)$ are available at a node X, calculate $\lambda_X(x)$.

- If all downstream messages $\pi_{U_i}(x)$ are available at a node X, calculate $\pi_X(x)$.

- If all messages except $\pi_{U_i X}(x)$ are available at node X, calculate $\lambda_{X U_i}(x)$ and send to node U_i.

- If all messages except $\pi_{Y_i X}(x)$ are available at node X, calculate $\pi_{X Y_i}(x)$ and send to node Y_i.

Stated more simply, each π and λ message is calculated as soon as the information on which it depends is available. This procedure results in messages first propagating inward from leaf nodes (and non-leaf evidence nodes), and then outward from the last node to be reached. Each message is calculated only once and retains its value until the end of the algorithm.

Approximate Loopy Belief Propagation
Many Bayesian inference problems are expressed in terms of graphs with loops or involve the computation of intractable integrals. This section covers several noteworthy approaches to applying BP when computing exact marginals is not possible.

Unlike tree-structured graphs, BP does not produce exact results on networks containing undirected loops. Intuitively speaking, evidence is passed around a loop, mistaken for new evidence, and "double-counted" [56]. In fact, nodes may not even converge toward a single value. Despite this disadvantage, "loopy BP" algorithms can perform very well. For instance, low-density parity check (LDPC) and turbo codes can achieve the maximum capacity of a channel using a decoding procedure which implements BP [57, 58].

In [58] an attempt is made to empirically study the limits of BP. This is done by running BP on several previously developed Bayesian networks containing multiple loops. BP is adapted to loopy graphs by simply updating every node in parallel on every iteration of the algorithm. The BP result is compared against that obtained by an importance sampling algorithm. Four networks are used: PYRAMID, a hierarchical network of binary nodes; toyQMR, a simulated medical diagnosis predictor initialized with random probabilities; ALARM, a network for detecting emergencies in intensive care patients, and QMR-DT, a binary network mapping 600 diseases to 4,000 findings (i.e., symptoms). In the cases of PYRAMID, toyQMR, and ALARM, BP converges to near the correct marginal probability. In the case of QMR-DT, however, BP does not converge and instead oscillates between two very different states. The authors are able to induce oscillation in the much smaller toyQMR network by making the prior probabilities of different diseases much lower, e.g., by choosing from the interval $[0, 0.1]$ rather than $[0, 1]$. Stability in the larger QMR-DT network can be induced by replacing the actual prior probabilities of disease by uniform random values on the interval $[0, 1]$. Adding a momentum term to the update reduces oscillation but does not necessarily improve the accuracy of the estimates for QMR-DT.

The success of turbo codes has prompted further analysis of BP, leading to additional performance guarantees. In [56] a belief revision procedure is described which partially corrects for the double-counting effect in networks consisting of a single loop of unknown nodes, each connected to a single evidence node. Under a set of easily satisfied restrictions, the belief revision procedure generates the correct MAP estimate. However, the distribution over the non-evidence nodes is not exact; double-counting of evidence leads to overconfident estimates.

In addition to loops, the problem of estimating marginal probability distributions is often intractable, or at least very computationally expensive. Approximate message passing algorithms attempt to solve this problem by instead estimating the distribution. Applied to CS, these algorithms are referred to as approximate message passing (AMP). Similar methods are known in other applications as "approximate BP" [59] or "relaxed BP" [60, 61]. In [62, 63] an i.i.d. Laplace (double-sided exponential) prior is placed on $\mathbf{x}$, and messages are approximated as Gaussian distributions. Under this assumption only a mean and variance must be passed, rather than a measure on the real line. The authors argue that the central limit theorem causes messages to converge toward Gaussian distributions as the problem size becomes large. Furthermore, the variances of all the messages are approximately identical. Expressions are derived for the large-system limit of the means $\mu_{\pi_{YX}}$ and μ_λ and variance σ_λ^2 of the messages. In vector notation, these are given as

$$\mu_{\pi_{xy}}^{t+1} = \eta\left(A^* \mu_{\lambda_{yx}}^t + \mu_{\pi_{xy}}^t ; \sigma_\lambda^{2^t}\right), \tag{2.26}$$

$$\mu_{\lambda_{yx}}^t = \mathbf{y} - A\mu_{\pi_{xy}}^t + \frac{1}{\delta}\mu_{\lambda_{yx}}^{t-1}\left\langle\eta'\left(A^* \mu_{\lambda_{yx}}^{t-1} + \mu_{\pi_{xy}}^{t-1} ; \sigma_\lambda^{2^{t-1}}\right)\right\rangle \tag{2.27}$$

$$\sigma_\lambda^{2^t} = \frac{\sigma_\lambda^{2^{t-1}}}{\delta}\left\langle\eta'\left(A^* \mu_{\lambda_{yx}}^{t-1} + \mu_{\pi_{xy}}^t ; \sigma_\lambda^{2^{t-1}}\right)\right\rangle, \tag{2.28}$$

where $\langle\cdot\rangle$ denotes the average of a vector. $\eta(x; b)$ and $\eta'(x; b)$ are the soft threshold function and its derivative, given as

$$\eta(x;b) = \text{sign}(x)(|x| - b)_+ \tag{2.29}$$

$$\eta'(x;b) = \frac{\partial^2}{\partial x \partial b}(\eta(x;b)). \tag{2.30}$$

Since the variance σ_λ^2 is identical for all π and all σ messages, it is not passed; only one scalar, representing the mean, is passed in each direction along the graph edges. This algorithm displays comparable performance to Basis Pursuit (Section 2.1) at much lower complexity. A similar algorithm for Basis Pursuit Denoising in AWGN as in (2.2) is also derived.

The work of [62] is extended and generalized in [53]. Arbitrary prior distributions $p(\mathbf{x})$ are allowed, instead of a Laplace distribution. Arbitrary measurement noise models are also considered, instead of a noiseless (BP) or AWGN (BPDN) distribution on $p(\mathbf{y}|\mathbf{x})$. Finally, message updates for a nonlinear sensing operator A are allowed by the algorithm, although little analysis is given for the nonlinear case.

Figure 2.3 shows a block diagram of the system model for GAMP. An unknown input vector $\mathbf{x} \in \mathbb{R}^N$ is generated based on a known $\mathbf{q} \in Q^N$. $\mathbf{x}$ is passed through a linear mixing channel with mixing operator A, generating the "noiseless" measurement matrix $\mathbf{z}$. Finally, $\mathbf{z}$ passes through an output channel, generating measurement vector $\mathbf{y}$. The input and output channels are separable, with mixing occuring only between $\mathbf{x}$ and $\mathbf{z}$. The conditional probability distributions are given by

$$p(\mathbf{x}|\mathbf{q}) = \prod_{n=1}^{N} p(x_n|q_n) \tag{2.31}$$

$$p(\mathbf{z}|\mathbf{x}) = \delta(\mathbf{z} - A\mathbf{x}) \tag{2.32}$$

$$p(\mathbf{y}|\mathbf{z}) = \prod_{m=1}^{M} p(y_m|z_m). \tag{2.33}$$

$\mathbf{q} \rightarrow \mathbf{x} \rightarrow \mathbf{z} \rightarrow \mathbf{y}$ forms a Markov chain.

The procedure for state updates in GAMP is given in Algorithm 2.1, while Figure 2.4 gives a graphical representation of the dependencies between the different variables. $\hat{\mathbf{x}}$ and $\hat{\mathbf{z}}$ are, as one would expect, the estimates of $\mathbf{x}$ and $\mathbf{z}$ at iteration t of the algorithm. The semantic meaning

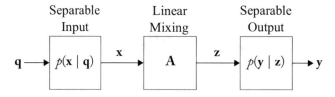

Figure 2.3: The GAMP system model. The measurement operation is divided into two separable but nonlinear measurement channels joined by a linear mixing operation.

of the other variables, however, is less obvious. $\hat{r}$ and $\hat{s}$ can be viewed as passing information "upstream," while $\hat{p}$ passes it downstream. In MMSE GAMP, discussed below, the τ variables may be interpreted as variances of the corresponding variables.

Algorithm 2.1 GAMP Algorithm.

Given $A \in \mathbb{R}^{M \times N}$, $\mathbf{y} \in \mathbb{R}^{M \times 1}$, $\mathbf{q} \, in \mathbb{R}^{N \times 1}$ and functions $g_{in}\left(r, q, \tau^{(r)}\right)$ and $g_{out}\left(p, y, \tau^{(}p)\right)$, produce a series of estimates $\hat{\mathbf{x}}(t)$ and $\hat{\mathbf{z}}(t)$.

Set $t = 0$, $\hat{s}(-1) = 0$

Set $\hat{\mathbf{x}}(0)$ and $\tau^{(x)}(0)$ to initial values

repeat

    for all $m \in [1, M]$ **do**

$$\tau_m^{(p)}(t) = \sum_{n=1}^{N} |a_{mn}|^2 \, \tau_n^{(x)}(t)$$

$$\hat{p}_m(t) = \sum_{n=1}^{N} a_{mn}\hat{x}_n(t) - \tau_m^{(p)}(t)\hat{s}_m(t-1)$$

$$\hat{z}_m(t) = \sum_{n=1}^{N} a_{mn}\hat{x}_m(t)$$

$$\hat{s}_m(t) = g_{out}\left(\hat{p}_m(t), y_m, \tau_m^{(p)}(t)\right)$$

$$\tau_m^{(s)} = -\frac{\partial}{\partial \hat{p}} g_{out}\left(\hat{p}_m(t), y_m, \tau_m^{(p)}(t)\right)$$

    end for

    for all $n \in [1, N]$ **do**

$$\tau_n^{(r)}(t) = 1/\sum_{m=1}^{M} |a_{mn}|^2 \, \tau_m^{(s)}(t)$$

$$\hat{r}_n(t) = \hat{x}_n(t) + \tau_j^{(r)}(t) \sum_{m=1}^{M} a_{mn}\hat{s}_m(t)$$

$$\hat{x}_n(t+1) = g_{in}\left(\hat{r}_n(t), q_n, \tau_n^{(r)}\right)$$

$$\tau_n^{(x)}(t+1) = \tau_n^{(r)}(t)\frac{\partial}{\partial \hat{r}} g_{in}\left(\hat{r}_n(t), q_n, \tau_n^{(r)}\right)$$

    end for

    $t = t + 1$

until t reaches the maximum number of iterations

The estimator functions g_{in} and g_{out} must also be chosen. [53] gives two methods for selecting these functions, based on MAP and minimum mean square error (MMSE) estimation.

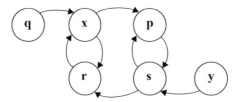

Figure 2.4: Dependencies of messages in the GAMP algorithm. Arrows entering a node indicate variables on which that node depends.

For MMSE estimation, the estimators are given by

$$g_{in}(\hat{r}, q, \tau_r) = \mathrm{E}\left(\hat{x}|\hat{r}, q, \tau^{(r)}\right) \tag{2.34}$$

$$g_{out}\left(\hat{p}, y, \tau^{(p)}\right) = \frac{1}{\tau^{(p)}}\left(\mathrm{E}\left(z|\hat{p}, y, \tau^{(p)}\right) - \hat{p}\right). \tag{2.35}$$

In the Expectation-Maximization Bernoulli-Gaussian Approximate Message Passing (EMBGAMP) algorithm [64], a Bernoulli-Gaussian prior is used, with prior PDF given by

$$p(\boldsymbol{\theta}) = \prod_{i=1}^{N}\left(q\delta(\theta_i) + (1-q)\frac{1}{\sqrt{2\pi\sigma_\theta^2}}\exp\left(\frac{\theta_i^2}{2\sigma_\theta^2}\right)\right). \tag{2.36}$$

That is, elements of $\boldsymbol{\theta}$ are assumed to take non-zero values with probability $1 - q$. The non-zero elements are then normally distributed with mean zero and variance σ_θ^2. α and σ_θ^2 unknown and are updated using expectation maximization on each iteration of the algorithm over the graph. In this way, the EMBGAMP algorithm is self-tuning and does not require the user to select any parameters to optimize its performance. It also is noteworthy in that it can outperform ℓ_1 basis pursuit for the noiseless case, discussed in Section 2.6.

In [65] the GAMP algorithm is extended to bilinear models, allowing the use of message passing as a solver for problems such as matrix completion, robust PCA, and dictionary learning. Under the bilinear model, both $A \in \mathbb{R}^{M \times N}$ and $X \in \mathbb{R}^{N \times L}$ are unknown and drawn independently from from their respective prior distributions. Note that X and Y are here allowed to be a matrix rather than a vector. The posterior PDF is given by

$$p(X, A, |Y) = p(Y|X, A)p(X)p(A) \tag{2.37}$$

$$= p(Y|AX)\,p(X)p(A). \tag{2.38}$$

An optional expectation maximization step may be also be included as in EMBGAMP [64], leading to the Expectation Maximization Bilinear Generalized Approximate Message Pasing (EM-BiG-AMP) algorithm. In Monte Carlo trials the EM-BiG-AMP algorithm shows performance near theoretical maximum for matrix completion problems.

2.5   VIDEO RECONSTRUCTION ALGORITHMS

Reconstruction of CS images is relatively straightforward, if computationally demanding. It is well known that natural images exhibit excellent energy compaction under the 2D discrete cosine transform (DCT) or any of several discrete wavelet transforms (DWT); this compressibility is exploited by the JPEG and JPEG2000 standards, respectively. Video compression algorithms, however, exploit both compressibility of individual frames and correlations between nearby frames. There are several examples in the literature of video reconstruction algorithms which incorporate this inter-frame information.

2.5.1   SPARSE AND LOW RANK DECOMPOSITION VIA COMPRESSIVE SENSING (SPARCS)

In [66], a quasi-static background with small moving targets is assumed. This assumption allows a video sequence to be represented as the sum of a low rank matrix, representing the background, and a sparse matrix, representing moving objects in the foreground. The SpaRCS algorithm, a greedy pursuit derived from the CoSAMP and ADMiRA algorithms [51], is then used to recover the video sequence. Figure 2.5 shows a conceptual block diagram of SpaRCS: on each iteration, CoSaMP and ADMiRA each attempt to solve for the other algorithm's residual. Although the SpaRCS algorithm performs well when the assumptions of static background and small motion are satisfied, it quickly fails when the background is not static or when targets are large. In addition, a relatively large number of frames (typically hundreds) are needed to accurately estimate the background. This is most likely a consequence of the algorithm's failure to take advantage of the known compressibility of the background image when performing reconstruction.

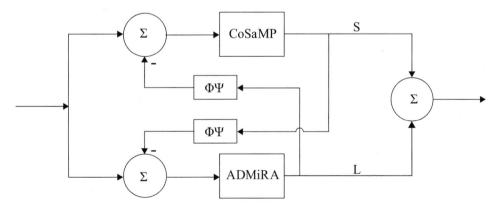

Figure 2.5: Conceptual block diagram of the SpaRCS algorithm.

2.5.2   CS MULTISCALE VIDEO RECOVERY (CS-MUVI)

The CS-MUVI algorithm [67] most nearly achieves the goal of fully utilizing known information, including both optical flow and compressibility of individual frames. In this algorithm, a special sensing matrix Φ is used which creates a well-conditioned matrix when combined with an upsampling operator U. The combined matrix ΦU is then used to reconstruct a low-resolution version of the video using least-squares estimation. Optical flow is then estimated using this low-resolution video and the resulting optical flow is used to improve reconstruction of the compressively sensed high-resolution video. This method, while highly effective, fails to reconstruct small objects since these objects are not visible in the low-resolution version of the video.

2.5.3   BACKGROUND SUBTRACTION

In [68], a static background with small targets is again assumed. In this case, however, the difference between two images is reconstructed. This is possible because as long as targets are small, the difference image is known to be spatially sparse. This algorithm achieves good results as long as all assumptions are met.

2.5.4   OTHER METHODS

In [69], global optical flow is estimated from compressive measurements of two video frames and then used to reconstruct both frames. While useful, this method requires the two input frames to be pure translations of one another. When optical flow is not constant across the entire image or is large (on the order of several pixels), the algorithm fails. This method also requires the use of a highly unorthodox sensing scheme in which compressively sensed pixels are compactly supported on the image. A sensing matrix of this type is likely to have undesirable properties.

In [70], dense optical flow is estimated from compressive measurements without performing any reconstruction. When the measurement rate is high, good estimates of optical flow are produced. However, an unconventional sensing scheme is again used, in which each measurement is compactly supported in one of the image's spatial dimensions. In fact, each compressive measurement is restricted to a single row of the uncompressed scene.

Although not strictly a CS reconstruction algorithm, the work of [71] deserves mention. The authors solve an ℓ_1 problem similar to (2.6) to estimate the optical flow field between stereoscopic images. Sparsity in the derivative of the optical flow field is promoted; this assumption holds true when motion is caused by a small number of rigid objects undergoing translational motion.

2.6   PERFORMANCE OF CS RECONSTRUCTION

Several approaches to the problem of CS reconstruction were in Section 2.4. Applicable guarantees and bounds on the performance of these algorithms are discussed here, as well as a comparison of the computational cost of the different algorithms.

2.6.1   COMPUTATION SPEED

Table 2.1 shows the time required to run each of the simulations in Figure 2.6. The comparison is imperfect, since the different algorithms are optimized to varying degrees, but trends are clear. Convex solvers are far more computationally demanding than greedy or message-passing algorithms.

Table 2.1: Runtimes required for phase transition plot

Algorithm	L1EQ	SPGL1	OMP	CoSaMP	GAMP	EMBGAMP
Runtime (hours)	51.5	9.0*	18.7	24.8	1.7*	2.2*

* SPGL1, GAMP, and EMBGAMP are partially or fully compiled C Code. Others are interpreted MATLAB.

2.6.2   CRAMER–RAO LOWER BOUND

The Cramer–Rao lower bound (CRLB) is arguably the most well-known bound on estimation problems. It is included here for completeness, although it is not very informative for the problem of CS. The CRLB gives a minimum bound on the variance of any unbiased estimator $\hat{\theta}$ of the vector θ given measurements $\mathbf{y}$. It is defined in terms of the Fisher information matrix $J(\theta)$ by the expression

$$\text{var}\,\hat{\theta} \geq J(\theta)^{-1} \tag{2.39}$$

$$J_{ij}(\theta) = -\underset{\mathbf{x}}{E}\left[\frac{\partial^2}{\partial\theta_i\,\partial\theta_j}\log(p(\mathbf{x}|\theta))\right]. \tag{2.40}$$

Equation (2.40) holds only under certain easily satisfied regularity conditions and is not the most universal definition of Fisher information. (2.39) holds only for unbiased estimators, and biased estimators may achieve lower variance. (It also does not take the prior distribution of θ into account, since $J(\theta)$ depends only on $p(\mathbf{y}|\theta)$). Allowing bias in estimation and defining a prior distribution $p(\theta)$ on the parameters to be estimated yields the Bayesian CRLB (BCRB), defined as, [72]:

$$E\left((\theta - \hat{\theta})(\theta - \hat{\theta})^T\right) \geq J_B^{-1}. \tag{2.41}$$

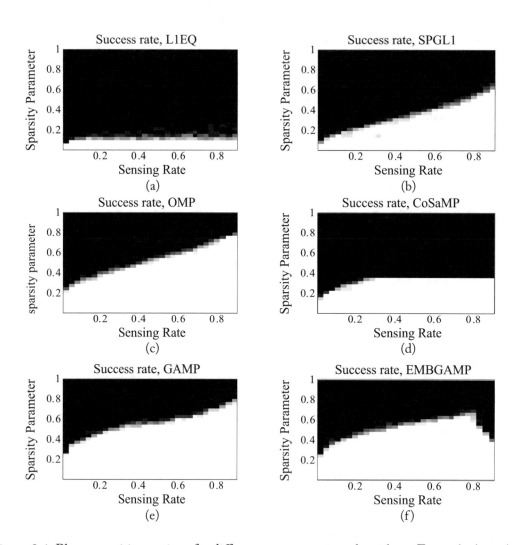

Figure 2.6: Phase transition regions for different reconstruction algorithms. For each algorithm, success rate is plotted as a function of sparsity ratio k/M and sensing rate r. Algorithms shown are (a) convex ℓ_1 BP using the l1-magic library's l1eq command, (b) ℓ_1 BP using SPGL1, (c) OMP, (d) CoSaMP, (e) GAMP with q, σ_θ^2 known, and (f) EMBGAMP.

J_B is the Bayesian Fisher information matrix and is expressed in terms of the standard Fisher information matrix J and an additional term reflecting the prior,

$$J_{ij} = \mathop{E}_{y,\theta}\left[-\frac{\partial^2 \log p(\mathbf{y},\boldsymbol{\theta})}{\partial\theta_i \partial\theta_j}\right] \tag{2.42}$$

$$= \mathop{E}_{\boldsymbol{\theta}}(J_{ij}) + \mathop{E}_{\boldsymbol{\theta}}\left[-\frac{\partial^2}{\partial\theta_i \partial\theta_j}\log p(\boldsymbol{\theta})\right] \tag{2.43}$$

$$= J_D + J_P, \tag{2.44}$$

where J_D and J_p are defined as

$$J_D = \mathop{E}_{\boldsymbol{\theta}}(J_{ij}) \tag{2.45}$$

$$J_P = \mathop{E}_{\boldsymbol{\theta}}\left[-\frac{\partial^2}{\partial\theta_i \partial\theta_j}\log p(\boldsymbol{\theta})\right]. \tag{2.46}$$

The BCRB for CS reconstruction problems is derived in [73] for a typical Bayesian CS problem. The measurement matrix Φ is selected with zero-mean i.i.d. elements of variance σ_Φ^2. Noise is AWGN with variance σ_w^2. $\boldsymbol{\theta}$ is distributed i.i.d. Bernoulli-Gaussian, with parameters q and σ_θ^2 and PDF $p(\theta_i) = q\delta(\theta_i) + (1-q)\frac{1}{\sqrt{2\pi\sigma_\theta^2}}\exp\left(\frac{\theta_i^2}{2\sigma_\theta^2}\right)$. Under these assumptions, the BCRB is calculated as

$$J_D = M\frac{\sigma_\Phi^2}{\sigma_w^2}\Psi^T\Psi \tag{2.47}$$

$$J_P = \frac{1-q}{\sigma_\theta^2}I \tag{2.48}$$

$$J_B = J_D + J_P = M\frac{\sigma_\Phi^2}{\sigma_w^2}\Psi^T\Psi + \frac{1-q}{\sigma_\theta^2}I. \tag{2.49}$$

Several problems with the bound in (2.49) are immediately evident. There is no choice of parameter values which creates a separation between regions where estimation is feasible or infeasible. Error actually increases as q increases (i.e., as the signal becomes more sparse). This is an artifact of the parametrization of $p(\boldsymbol{\theta})$, which leads to increasing signal energy as more non-zero components are added to the signal. The largest problem, though, is that there is no dependence on the number of parameters N, only on the number of measurements M. In the large system limit the J_D term dominates and drives the minimum error to zero. Likewise, in the noiseless ($\sigma_w^2 = 0$) case, the J_D term is infinite. The BCRB therefore reduces to the trivial lower bound of $E[(\boldsymbol{\theta} - \hat{\boldsymbol{\theta}})^T(\boldsymbol{\theta} - \hat{\boldsymbol{\theta}})] \geq 0$ for both the noiseless case and the large-system limit.

2.6.3   PHASE TRANSITION

A more productive alternative to the BCRB of Section 2.6.2 is clearly needed. One way to characterize the limits on reconstruction performance is to examine an algorithm's phase transition, loosely defined as the dividing line between regions where reconstruction is good vs. poor. In this section, theoretical and empirical results on the phase transition regions of several CS reconstruction algorithms are presented and discussed.

Donoho and Tanner [74, 75] give a limit on the feasible region for ℓ_1 basis pursuit. The case of a random Gaussian measurement matrix $A \in \mathbb{R}^{M \times N}$ is considered. As the problem size approaches infinity, ℓ_1 basis pursuit generates a correct solution with probability approaching 1 as long as the condition

$$k < M \left(2e \log(N/M)\right)^{-1} \tag{2.50}$$

is met. As usual, $k = \|\theta\|_0$ is the sparsity level of the signal to be reconstructed. Empirical evidence also indicates that this limit holds for finite problems of reasonable size. Figure 2.6 shows the average reconstruction success rate over 10 trials, for a range of sparsity levels and sensing rates. Number of measurements $M = 1024$ was held constant and N and k were varied to achieve the desired ratios. Success rate is defined here as the percentage of trials with $\left\| \theta - \hat{\theta} \right\|_\infty < 10^{-3}$. Although the phase transition of (2.50) is derived for i.i.d. Gaussian A, in [75] it is shown that many other choices of A show an identical phase transition, including Bernoulli, random Fourier, and random Hadamard matrices.

Additional refinements have been made to the Donoho-Tanner phase transition. It is known [76] that for exactly known sparsity level $\|\theta\|_0 = k$, the maximum k grows proportionally to sensing rate r rather than $(1/r \log(1/r))^{-1}$ as in (2.50). In [76], however, (2.50) is shown to be tight for the ℓ_1 problem, in agreement with [74]. This indicates a hard limit for the ℓ_1 basis pursuit problem; any improvements must come from other methods.

Different reconstruction algorithms may be compared by evaluating their phase transition region, as well as execution speed and error rates on finite-length data. Figure 2.6 shows empirically estimated phase transitions for two convex solvers, L1EQ from the l1-magic library [77] and SPGL1 [46]; two greedy algorithms, OMP [47] and CoSaMP [51]; and two message passing algorithms, GAMP[53] and EMBGAMP [64]. L1EQ, when run with the default parameters, showed very poor performance; this demonstrates the need for appropriate algorithm tuning. SPGL1, which nominally solves an equivalent problem, showed much better performance. OMP and CoSaMP show comparable performance to convex solvers at low sensing rates. The chosen CoSaMP implementation, however, requires $k < M/3$ for the successful solution of an internal least-squares problem. This leads to suboptimal performance at higher sensing rates. The message-passing algorithms both perform extremely well, with EMBGAMP slightly outperforming GAMP at low sensing rates but suffering from stability problems as more measurements are taken.

2.6.4   SPARSE RECOVERY IN NOISE

All the phase transitions discussed above are for the case of noiseless measurement and perfectly sparse $\boldsymbol{\theta}$. A natural question is whether CS reconstruction algorithms are robust. How much noise can be added before the support of $\boldsymbol{\theta}$ can no longer be reliably determined? For the question of noisy CS reconstruction, [78] introduces the concept of the noise-sensitivity phase transition.

The noise-sensitivity phase transition is defined as follows. Let $\mathbf{y}$ be determined as in (2.2), as usual, under AWGN with $\mathbf{w} \sim \mathcal{N}(\mathbf{0}, \sigma_w^2 \mathbf{I})$. Let $MSE = \|\hat{\boldsymbol{\theta}} - \boldsymbol{\theta}\|_2^2 / N$ be the mean squared reconstruction error. The ratio MSE/σ_w^2 is defined as the *noise sensitivity*. The authors then define a worst-case noise sensitivity for BPDN reconstruction as a function of sensing rate r and sparsity ratio k/M. This worst-case noise sensitivity is bounded over exactly the same region as the phase transition originally given in (2.50). This implies that BPDN is robust to noise and shows graceful degradation as noise increases.

2.7   RECOVERY OF COMPRESSIBLE SIGNALS

Naturally occurring signals are generally compressible but not perfectly sparse. For $\boldsymbol{\theta}$ following a power law (for instance), how well must energy be concentrated before data can be approximated as sparse? In this section, simulations are performed to characterize reconstruction for varying levels of signal compressibility, and the results are discussed. Approaches from the literature are described and another bound is put forward based on the noise sensitivity phase transition of [78].

In order to explore the phase transition for power law distributed data, $\boldsymbol{\theta}$ was drawn as a symmetric α-stable distribution with stability parameter α_θ, scale parameter $\gamma_\theta = 1$, and center parameter $\mu_\theta = 0$. The PDF of θ is not analytically expressible, but its characteristic function $\phi_\theta(t)$ is given by

$$\phi_\theta(t) = \exp\left(jt\mu - |\gamma_\theta t|^{\alpha_\theta}\right). \tag{2.51}$$

The stability parameter $\alpha_\theta \in (0, 2]$ controls the tail behavior, and therefore the compressibility, of the signal. $\alpha_\theta = 2$ corresponds to the Gaussian distribution, while $\alpha_\theta = 1$ is the Cauchy distribution. Except for the Gaussian ($\alpha_\theta = 2$) case, all α-stable distributions have power-law tails with PDF proportional to $|x|^{-(1+\alpha)}$ for large $|x|$. θ has infinite variance, but was normalized in simulations so that $\|\boldsymbol{\theta}\|_2 = 1$ to avoid numerical overflow problems at small α.

Figure 2.7 explores the phase transition for power-law distributed $\boldsymbol{\theta}$. Unlike the k-sparse case, the phase transition region does not show a sharp boundary between successful and unsuccessful reconstructions. There also appears to be much less benefit from increased sensing rates. Instead, a steady decrease in success rate for increasing α is observed.

A natural question arises: what is occurring in the region of perfect (or near-perfect) success rates? Is this a true phase transition or does it instead depend on the error tolerance, set somewhat arbitrarily at 10^{-3}? To answer this question, the Modulation error ratio (MER) of each method is displayed in Figure 2.8. MER is the ratio of signal energy to error energy and is

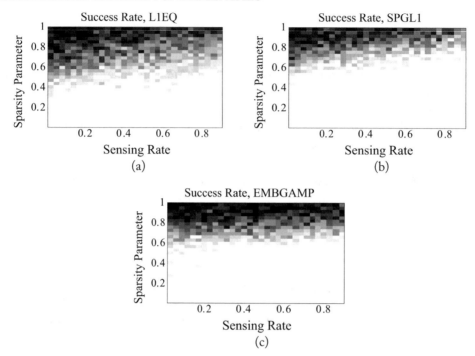

Figure 2.7: Phase transitions for symmetric stable distributed $\boldsymbol{\theta}$. Sparsity is controlled by stability parameter α.

defined as,

$$MER = \frac{\|\boldsymbol{\theta}\|_2^2}{\left\|\boldsymbol{\theta} - \hat{\boldsymbol{\theta}}\right\|_2^2}. \tag{2.52}$$

Examining the MER, there is again no clear line between success and failure of the reconstruction; instead there is a steady decrease in MER as α is increased (i.e., as the signal becomes less sparse). In order to compare the performance of different algorithms, Figure 2.9 shows the difference in MER between EMBGAMP and SPGL1. The plotted quantity is MER under EMBGAMP minus MER under SPGL1. A large (> 25 dB) advantage for EMBGAMP is seen for small α (very sparse $\boldsymbol{\theta}$). SPGL1, on the other hand, outperforms EMBGAMP for cases of larger α and higher sensing rates.

The disappearance of the phase transition in power-law data is worrying. If guarantees of reconstruction performance only apply to perfectly sparse data, they are of little use in most potential applications. Fortunately, there are bounds on reconstruction of imperfectly sparse data. The first approach is to treat all but the k largest components of $\boldsymbol{\theta}$ as noise; using this approach, bounds have been derived for ℓ_1 reconstruction [9] and CoSaMP [51].

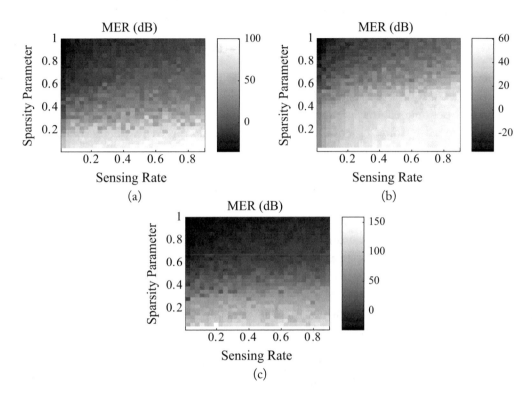

Figure 2.8: Modulation error ratio for symmetric stable distributed $\boldsymbol{\theta}$. Sparsity is controlled by stability parameter α (plotted on y-axis). (a) L1EQ, (b) SPGL1, and (c) EMBGAMP.

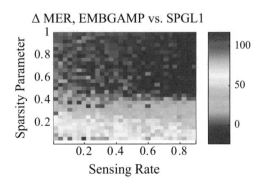

Figure 2.9: Comparison of MER of EMBGAMP vs. SPGL1.

Let $\boldsymbol{\theta}$ be the (not necessarily sparse) signal to be reconstructed, and let $\boldsymbol{\theta}_k$ be the vector consisting of the k largest entries of $\boldsymbol{\theta}$, with zeros elsewhere. In [9] it is shown that the ℓ_1 BP solution $\hat{\boldsymbol{\theta}}$ satisfies

$$\left\| \hat{\boldsymbol{\theta}} - \boldsymbol{\theta} \right\|_1 \leq C_0 \left\| \boldsymbol{\theta} - \boldsymbol{\theta}_s \right\|_1, \tag{2.53}$$

as long as the restricted isometry property of (2.4) is satisfied with isometry constant $\delta_{2k} < \sqrt{2} - 1$. C_0 is a constant, given explicitly in [9].

Another possible approach is inspired by the noise-sensitivity phase transition [78]. Assume that there exists some function $g(r, \rho)$ which bounds the MSE of a reconstruction on sparse data. That is,

$$\frac{\left\| \boldsymbol{\theta} - \hat{\boldsymbol{\theta}} \right\|_2^2}{N} \leq \sigma_w^2 g(r, \rho), \quad \|\boldsymbol{\theta}\|_0 \leq k \tag{2.54}$$

in the large-system limit as $M \to \infty$, $M/N \to r$, and $k/M \to \rho$. [78] gives strong evidence that such a $g(r, \rho)$ exists, and that it is finite over the same region as the Donoho-Tanner phase transition. Now, let $\boldsymbol{\theta}$ be non-sparse, but instead distributed i.i.d. symmetric, and centered on 0, with PDF $p_\theta(\theta)$ and CDF $F_\theta(\theta)$.

Proceed by dividing $\boldsymbol{\theta}$ into a "signal" component $\boldsymbol{\theta}_s$, which can be recovered with bounded error, and a "noise" component $\boldsymbol{\theta}_n$, which we abandon hope of recovering. Setting the threshold of signal and noise at $F_\theta^{-1}(\rho r/2)$, we have

$$\theta_{n,i} = \begin{cases} \theta_i, & |\theta_i| \leq -F_\theta^{-1}(\rho r/2) \\ 0, & \text{otherwise} \end{cases} \tag{2.55}$$

$$\boldsymbol{\theta}_s = \boldsymbol{\theta} - \boldsymbol{\theta}_n. \tag{2.56}$$

That is, $\boldsymbol{\theta}_n$ contains all of the signal components with whose magnitude is less than the top ρr quantile, with zeros elsewhere. Likewise, $\boldsymbol{\theta}_s$ contains all signal components with magnitude in at least the top ρr quantile. We can now write the CS sensing model as

$$\mathbf{y} = \mathbf{A}\boldsymbol{\theta} = \mathbf{A}\boldsymbol{\theta}_s + \mathbf{A}\boldsymbol{\theta}_n = \mathbf{A}\boldsymbol{\theta}_s + \tilde{\mathbf{w}}. \tag{2.57}$$

In the above, $\tilde{\mathbf{w}} = \mathbf{A}\boldsymbol{\theta}_n$ represents the effect of $\boldsymbol{\theta}_n$ on the measurement. By the central limit theorem,

$$\tilde{\mathbf{w}} \sim \mathcal{N}\left(\mathbf{0}, \sigma_{\theta_n}^2 \mathbf{A}\mathbf{A}^T\right), \tag{2.58}$$

where

$$\sigma_{\theta_n}(r, \rho) = \int_{F_\theta^{-1}(\rho r/2)}^{-F_\theta^{-1}(\rho r/2)} \theta^2 p_\theta(\theta) d\theta \tag{2.59}$$

is the variance of each element of $\boldsymbol{\theta}_n$. Most commonly used sensing matrices, including deterministic and random orthoprojectors and i.i.d. Gaussian matrices, satisfy $\mathbf{A}\mathbf{A}^T \to \mathbf{I}$ in the large-system limit. In this case, the assumption of AWGN holds. We can now bound the MSE as

$$\frac{\left\| \hat{\boldsymbol{\theta}} - \boldsymbol{\theta}_s \right\|_2^2}{N} \leq \sigma_{\theta_n}^2 (r, \rho) g(r, \rho). \tag{2.60}$$

Thus, the theory of compressed sensing is not only useful for perfectly sparse signals, it also bounds the error of CS reconstruction algorithms when operating on compressible signals which more closely model those found in nature.

2.8   DEEP LEARNING FOR COMPRESSED SENSING RECONSTRUCTION

In this section, we discuss recent advances in deep learning for CS reconstruction, which have been outperforming classical solvers in both performance and computational efficiency. We survey recent advances in the field, and discuss how their network structure is designed to improve the CS reconstruction.

2.8.1   CONVOLUTIONAL NEURAL NETWORKS

Before we discuss deep learning for CS reconstruction, we will introduce convolutional neural networks, the modern workhorse of computer vision processing that is used. Convolutional neural networks (CNNs) have risen to prominence in computer vision after showing competitive or best-in-class performance on a wide variety of difficult problems. The CNN may be viewed as a feedforward neural network which uses several tricks to exploit prior knowledge of the behavior of many naturally occurring signals, including digital images. In this section, a basic overview of the structure of CNNs, their training, and several recent CNN-based algorithms for Compressive Sensing Reconstruction have been discussed.

CNNs mitigate the problem of overfitting by weight-sharing and max-pooling, both of which reduce the number of parameters in the computational model while still modeling complex input-output relationships. Figure 2.10 gives a 1D example of a small linear convolutional layer. Each hidden-layer neuron is connected to only a small neighborhood of the input layer, in this case 3 samples wide. Already, the number of weights has been reduced from DP_1 to $3P_1$. The weights are further shared between neurons, so that every node performs the same inner product on its neighborhood; this weight sharing is shown in Figure 2.10 by the colors of the arrows. The number of model parameters is now only 3. In addition to the massive reduction in model parameters, the P_1 inner products, each requiring D multiplications, are replaced by a single convolution operation. With the use of the FFT, the complexity of this operation has been reduced to $D \log D$.

Convolution is, of course, a linear operation; nonlinearity must be added to the network. In a typical CNN this is accomplished via activations and max-pooling, again shown in Figure 2.10. In max-pooling, an input image's pixels are simply binned into a grid, and the maximum of

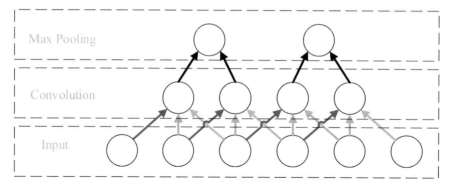

Figure 2.10: Convolutional neural network architecture.

each bin is passed to the next layer. In this way, max-pooling functions as a sort of nonlinear downsampling.

2.8.2   RBM FOR COMPRESSED SENSING RECONSTRUCTION

Machine learning, and particularly deep learning, may be used to reconstruct compressively sensed images with high performance. In [79, 80], a restricted Boltzmann machine (RBM) is used as a prior for CS reconstruction. The case of sparse $\mathbf{x}$ (i.e., $\Psi = \mathrm{I}$) is considered, but results can easily be extended to the case of dense $\mathbf{x}$ and sparse $\boldsymbol{\theta}$. This reconstruction method is tested on the MNIST digit data and shows performance near that of a "support oracle" algorithm which is given the support of the signal. This is possible because the RBM is generative and attempts to learn the probability distribution of the training data. A 2-layer RBM consisting of binary nodes is joined to a Bernoulli-Gaussian prior on $\mathbf{x}$, which is then reconstructed with AMP. In this way, the algorithm can be viewed as a more sophisticated form of re-weighted CS reconstruction algorithms [81] or Model-based CS [16].

The system model consists of a Markov chain $\mathbf{h} \to \mathbf{v} \to \mathbf{x} \to \mathbf{y}$ with conditional probabilities given as

$$\mathbf{h}|\mathbf{v} \sim \mathrm{Bern}\left(\sigma\left(\mathrm{W}^{(1)^T}\mathbf{v}\right)\right) \tag{2.61}$$

$$\mathbf{v}|\mathbf{h} \sim \mathrm{Bern}\left(\sigma\left(\mathrm{W}^{(1)}\mathbf{h}\right)\right) \tag{2.62}$$

$$p(x_i|v_i) = \begin{cases} \delta(x_i), & v_i = 0 \\ \frac{1}{\sqrt{2\pi\sigma_x^2}}\exp\left[\frac{x_i^2}{2\sigma_x^2}\right], & v_i = 1 \end{cases} \tag{2.63}$$

$$\mathbf{y}|\mathbf{x} \sim \mathcal{N}\left(\Phi\mathbf{x}, \sigma_w^2\mathrm{I}\right). \tag{2.64}$$

That is, the hidden layer $\mathbf{h}$ and visible layer $\mathbf{v}$ are related as in the binary RBM described by (3.2), while the unknown data vector $\mathbf{x}$ and measurement vector $\mathbf{y}$ are related by the standard

CS sensing operation in AWGN. $\mathbf{x}$ is independently but not identically distributed according to a BG prior, taking non-zero values with element-wise probability given by $\Pr(x_i \neq 0) = \Pr(v_i = 1)$.

2.8.3 STACKED DENOISING AUTOENCODER FOR CS RECONSTRUCTION

In [82], a Stacked Denoising Autoencoder (SDA) architecture is presented for CS reconstruction. The work presents two cases wherein linear and nonlinear measurement paradigms are considered. A three-layer SDA is developed such that each layer has an input size corresponding to the original signal and output size corresponding to the measurement vector or vice versa. In the nonlinear measurement paradigm, one more layer is added to SDA architecture which maps the original signal to its measurement to aid the measurement matrix to become used to input class of signals. In both paradigms, mean squared error is used as a loss function for the training set as shown in Eqs. (2.65) and (2.66), respectively:

$$\mathcal{L}(\Omega_L) = 1/1 \sum_{i=1}^{1} \left\| \mathcal{M}_L \left(\mathbf{y}^{(i)}, \Omega_L \right) - \mathbf{x}^{(i)} \right\|_2^2, \tag{2.65}$$

where $\mathcal{M}_L(\mathbf{y}^{(i)}, \Omega_L)$ represents nonlinear mapping to obtain the signal estimate.

$$\mathcal{L}(\Omega_{NL}) = 1/1 \sum_{i=1}^{1} \left\| \mathcal{M}_{NL} \left(\mathbf{y}^{(i)}, \Omega_{NL} \right) - \mathbf{x}^{(i)} \right\|_2^2. \tag{2.66}$$

To solve the issue of increasing computational complexity due to increase in signal size, a signal is divided into overlapping or non-overlapping blocks and hence a blocky measurement matrix is used in [82]. This fails to utilize the interdependence between different block reconstructions which is addressed in [83].

2.8.4 DEEPINVERSE AND DEEPCODEC

The DeepInverse architecture developed in [83] consists of one fully connected layer followed by three convolutional layers. The architecture takes compressed sensing measurements, i.e., y as input and uses full connected layer to increase the dimensionality and obtain an estimate of original signal.

$$\widetilde{\mathbf{x}} = \boldsymbol{\Phi}^T y, \tag{2.67}$$

wherein $\boldsymbol{\Phi}^T$ is implemented using weights of fully connected layer. A nonlinear mapping is learned by the architecture to obtain the original signal from signal estimate obtained using the fully connected layer. DeepCodec architecture developed in [84] unlike DeepInverse utilizes nonlinear undersampled measurements to recover original signal. The Deepcodec applies the following two steps.

1. It learns the mapping from the original signal $\mathbf{x} \in \mathbb{R}^N$ to undersampled measurements $\mathbf{y} \in \mathbb{R}^M$. The first layer changes the input size from $N \times 1 \times 1$ to $M \times 1 \times r$ wherein $r = N/M$ implying reduction in dimension of output signal with increase in number of channels by the same amount. In order to obtain a measurement signal of size $M \times 1 \times 1$ several convolutional layers are applied. Now in Step 2, the reconstruction is performed in measurement domain.

2. In order to decode or increase the signal dimensionality, several convolutional layers increase the signal dimension from $M \times 1 \times 1$ to $M \times 1 \times r$ followed by a sub-pixel convolution layer which produces an output of size $N \times 1 \times 1$.

The computational complexity of the CS recovery method presented in [84] is less expensive compared to [83] while obtaining the better signal recoveries.

2.8.5   IMAGE RECOVERY USING RIDE

[85] enforces Recurrent Image Density Estimator (RIDE) as image prior and obtains better compressive image reconstructions. The authors use 2D Spatial Long Short Term Memory units to capture long term dependencies. They use MAP principle to find the desired image as shown in Eq. (2.68):

$$\hat{\mathbf{x}} = \arg \max_{\mathbf{x}} \Pr(\mathbf{x} \mid \mathbf{y}) = \arg \max_{\mathbf{x}} \Pr(\mathbf{x}) \Pr(\mathbf{y} \mid \mathbf{x}), \tag{2.68}$$

where

$$\Pr(x) = \prod_{ij} \Pr\left(x_{ij} \mid \mathbf{h}_{ij}, \boldsymbol{\theta}\right) \tag{2.69}$$

$$\Pr\left(x_{ij} \mid \mathbf{h}_{ij}, \boldsymbol{\theta}\right) = \sum_{c,s} \Pr\left(c, s \mid \mathbf{h}_{ij}, \boldsymbol{\theta}\right) \Pr\left(x_{ij} \mid \mathbf{h}_{ij}, c, s, \boldsymbol{\theta}\right) \tag{2.70}$$

likelihood is given by

$$\Pr(\mathbf{y} \mid \mathbf{x}) \propto \exp\left(-\frac{\|\mathbf{y} - A\mathbf{x}\|^2}{\sigma^2}\right), \tag{2.71}$$

$\hat{\mathbf{x}}$ represents desired image, x_{ij} represents pixel intensity at location ij in image x, and $\mathbf{h}_{ij}$ represents the hidden representation which gives the causal context around the pixel at location ij. Log posterior obtained from Eq. (2.68) is maximized using gradient ascent with a momentum parameter. The method is tested on BSDS300 test set which is cropped to size 160×160. Results show that the algorithm performs better than D-AMP and TVAL3 in terms of Peak SNR (PSNR) and Structural Similarity Index metrics (SSIM).

2.8.6 CS RECOVERY USING PIXELCNN++ AND MAP INFERENCE

In [86], authors use a deep autoregressive model PixelCNN++ as the image prior to obtain better reconstruction quality. A deep CNN with residual connections is used to model context pixels $x_{<i}$. A conditional distribution model where parameters depend on context is learned using maximum likelihood training. Backpropagation to inputs is then used to obtain gradients of the density with respect to the image. To set up an optimization problem in the case of a single-pixel camera with no measurement noise, a hard constraint method is used.

$$\hat{X} = \arg\max_{\mathbf{x}} \log\left(\Pr_{\theta}(\mathbf{X})\right)$$

$$\text{s.t. } C_1 = \{\mathbf{X} : \mathbf{Y} = f(\mathbf{X})\} \text{ and}$$

$$C_2 = \{\mathbf{X} : 0 \leq \mathbf{X}_{ij} \leq 1 \; \forall i, j\}.$$

Projected Gradient descent is used to solve this optimization problem and the solution in single-pixel camera case is given by

$$\mathbf{j_k} = \mathbf{h}_k - \Phi^T \left(\Phi\Phi^T\right)^{-1} \left(\Phi\mathbf{h}_k - \mathbf{y}\right), \tag{2.72}$$

where $\mathbf{j_k}$ and $\mathbf{h}_k$ are vector representations of matrices $\mathbf{J}_k$ and $\mathbf{H}_k$, respectively, and

$$\mathbf{H}_k = \mathbf{X}_k + \alpha \nabla_{\mathbf{X}} \log\left(\Pr_{\theta}(\mathbf{X}_k)\right)$$

$$\mathbf{J}_k = \prod_{C1}(\mathbf{H}_k)$$

$$\mathbf{X}_{k+1} = \prod_{C2}(\mathbf{J}_k).$$

They show that in presence of Gaussian distributed noise, the MAP estimation can be written as

$$\hat{X} = \arg\max_{\mathbf{x}} \log\left(\Pr_{\theta}(\mathbf{X}) + \lambda \|\mathbf{Y} - f(\mathbf{X})\|^2\right)$$

and gradient descent over likelihood can be replaced to

$$\mathbf{J}_k = \mathbf{H}_k - \alpha f'(\mathbf{H}_k)(\mathbf{Y} - f(\mathbf{H}_k)).$$

To avoid all neighboring pixels getting the same value due to image prior, they apply pixel dropout and gradient update is replaced by stochastic gradient update given by

$$\mathbf{H}_k = \mathbf{X}_k - \alpha \mathbf{M} \circ \nabla_{\mathbf{X}} \log\left(\Pr_{\theta}(\mathbf{X}_k)\right), \tag{2.73}$$

where **M** is a binary mask determined by pixel dropout ratio. The initial image is sampled from uniform random distribution followed by image splitting into patches of size 64×64. The prior gradient update is performed followed by patch stitching. The likelihood step is performed on image of original size. The presented approach performs better in terms of PSNR and SSIM compared to TVAL3 and OneNet for simulated SPC reconstruction with images of size 128×128 and 256×256 at measurement rate 10%, 25% and 5%, 10%, respectively. The method works well in case of Real SPC Reconstruction at 30% and 15% measurement rates compared to TVAL3 and RIDE-CS.

2.8.7   THE RECONNET ARCHITECTURE

Another architecture ReconNet is presented in [87] provides benefits in terms of reconstruction speed as well as quality over different measurement rates. The method involves obtaining image patches of size 33×33 from multiple images and obtaining their CS measurements using a row orthonormalized random Gaussian matrix. The original signal is the luminance component of image block and a measurement vector is obtained by product of original signal and the Gaussian matrix. This CS measurement vector along with the original signal serve as the input and label, respectively.

The measurement vector is given as an input to a fully connected layer to obtain a 2D array of size 33×33 which the initial estimate of the original signal. This estimate is passed through a ReconNet unit which has three convolutional layers with kernel size 11×11, 1×1, and 7×7 wherein the last layer produces only 1 feature map which serves as the output of the network. An entire image is reconstructed using image patches obtained from the network and passed through a BM3D denoiser to avoid artifacts due to block processing.

Different loss functions are explored while training the ReconNet architecture including *Euclidean* loss and *Euclidean* + *Adversarial* loss. The Euclidean loss is shown in Eq. (2.74):

$$L(\Theta) = 1/B \sum_{i=1}^{B} \| f(\mathbf{y_i}, \Theta) - \mathbf{x_i} \|^2 \,, \tag{2.74}$$

where Θ represents network parameters, B represents total number of image blocks in one batch of the training set, $f(\mathbf{y_i}, \Theta)$ represents output obtained from the network, and x_i is the ith input patch. It was shown empirically that two ReconNet units work well to produce good reconstruction quality when trained using this loss function.

The loss equation for ReconNet with *Euclidean* + *Adversarial* loss includes a Generator and Discriminator loss. The Discriminator loss is shown in Eq. (2.75):

$$L_D = 1/B \sum_{i=1}^{B} (L_{CE}(D(\mathbf{x_i}), 1) + L_{CE}(D(G(\mathbf{y_i})), 0)), \tag{2.75}$$

where L_{CE} represents the cross-entropy loss, $L_{CE}(D(\mathbf{x_i}), 1)$ represents the ability of Discriminator to classify real images, and $L_{CE}(D(G(\mathbf{y_i})), 0)$ represents the ability of Discriminator to classify fake images. The Generator Loss is shown in Eq. (2.76):

$$L_G = \lambda_{rec}/B \sum_{i=1}^{B} \|G(\mathbf{y_i}) - \mathbf{x_i}\|^2 + \lambda_{adv}/B \sum_{i=1}^{B} (L_{CE}(D(G(\mathbf{y_i})), 1), \tag{2.76}$$

which is the sum of Euclidean and Adversarial loss. The G, i.e., Generator network uses only one ReconNet unit and is fixed and D, i.e., Discriminator uses three convolutional layers followed by a fully connected layer to obtain a single probability value.

The network is trained for four different measurement rates, i.e., 0.25, 0.10, 0.04, and 0.01. The results show that at low measurement rates of 0.10, 0.04, and 0.01 both version of ReconNet provide better reconstructions than their counterparts. At higher measurement rates of 0.25 and 0.10, the second variant of ReconNet with Euclidean and Adversarial loss performs better than ReconNet(Euc) in terms of PSNR and Visual Quality while at low measurement rates ReconNet(Euc) performs better in terms of PSNR however visually reconstructions using ReconNet(Euc + Adv) are more sharp and preserve finer details.

Further variants of these architectures are trained wherein a fully connected layer is added in the beginning to learn the measurement matrix ϕ. Significant gain is observed in terms of PSNR using both the Euclidean loss variant as well as Euclidean + Adversarial variant. Furthermore, fully connected layers are replaced with circulant layers which reduces the number of parameters and hence, computational complexity at the cost of minimal decrease in PSNR.

2.9   SUMMARY

In this chapter, the theory of compressed sensing was introduced. Hardware implementations of CS sensors, the single-pixel camera and magnetic resonance imaging, were discussed. The convex, greedy, and message-passing reconstruction algorithm families were discussed, and several specific examples were elaborated on. Specialized reconstruction algorithms for image and video data were reviewed [88]. The performance of CS signal recovery algorithms was discussed through both Monte Carlo simulations and analytically derived bounds. Finally, state-of-the-art deep learning approaches for CS recovery were presented.

CHAPTER 3

Computer Vision and Image Processing for Surveillance Applications

Reconstruction-free compressive vision depends heavily on existing work in the fields of traditional image and video processing, and on recent advances in machine learning and computer vision. The line between image processing and computer vision is sometimes difficult to define, as the image processing community adopts machine learning and artificial intelligence approaches to the problems of the field. However, an overview of all of computer vision and image processing is out of scope for this book, and we refer the readers to more comprehensive references for computer vision [89] and image processing [90]. In this chapter, we focus on surveillance applications, particularly tools related to classification, detection, and tracking that we will utilize in the subsequent chapter for their compressive analogs.

3.1 SURVEILLANCE OVERVIEW

Modern-day surveillance relies heavily on the dense proliferation and usage of surveillance cameras. This includes the CCTV networks in the United Kingdom, police cameras, as well as security cameras for buildings and facilities. While traditionally, surveillance has required a human operator or analyst to view the incoming images/video and make decisions, the field of image processing and computer vision has enabled semi-automated or automated tools for these purposes. Most of the initial successes have come in the areas of face detection/recognition [91], and the tracking and analysis of human activity [92]. A main area of focus has been implementing real-time algorithms for surveillance [93], and the computational resources needed for such systems. In this book, we propose reconstruction-free computer vision as a low-cost alternative to traditional algorithms. In the next sections of this chapter, we survey tools that will be useful for proposed algorithms in Chapter 4.

3.2 CLASSIFICATION AND DETECTION ALGORITHMS

Classifying the content of images is a classic problem in computer vision and forms an essential part of many other problems. Our primary interest is in the two-class problem of target detection and its application to tracking algorithms. This section covers a selection of classification algorithms, approximately ordered from the simplest and earliest to most recent and most complex.

3.2.1 MACH FILTER

The maximum average correlation height (MACH) filter is a well-known approach to offline training of a correlation-based filter to optimally detect a single object in video[94, 95]. A MACH filter is trained using examples from the positive (target) and negative (noise or background) classes. In its most basic form the MACH filter is defined (in the frequency domain) by,

$$\mathbf{H} = \left(D_y + S_x\right)^{-1} \mu_x, \tag{3.1}$$

where D_y is the power spectral density of the noise examples and S_x is the similarity matrix of the true examples. S_x and D_y are diagonal matrices. μ_x is the mean of the target examples. Choosing a filter $\mathbf{h}$ as in (3.1) maximizes the average correlation height of the detection peaks while minimizing the average similarity measure (ASM) between the targets and noise [96]. Several modifications and variations on the MACH filter exist, including, the optimal trade-off MACH (OT-MACH) filter [97, 98], which also optimizes output noise variance and average correlation energy.

3.2.2 DEEP BOLTZMANN MACHINES

The deep Boltzmann machine (DBM) is a network-based bio-inspired computational model. Unlike traditional neural networks, Boltzmann machines are undirected networks and can be run in reverse to generate example data. They are also stochastic: a node's inputs do not define its output but rather a probability distribution over its possible outputs. The DBM expands on the Boltzmann machine concept by adding an arbitrarily large number of layers to the network and defining an unsupervised layer-by-layer pre-training step. Figure 3.1 gives a visual representation of a DBM. A brief sketch of DBM evaluation and training is given here; we follow the conventions and definitions of [99] and invite readers to explore this work for a more complete treatment of the subject.

A standard single-layer restricted Boltzmann machine (RBM) consists of a visible layer $\mathbf{v} \in \{0, 1\}^{D \times 1}$ and a hidden layer $\mathbf{h} \in \{0, 1\}^{P \times 1}$ of nodes. The values of the nodes in each layer are conditionally independent given the value of the other layer, with conditional probabilities given by

$$p\left(h_j = 1 | \mathbf{v}, \mathbf{h}_{-j}\right) = \sigma\left(\sum_i^D w_{ij} v_i\right) \tag{3.2}$$

$$p\left(v_i = 1 | \mathbf{v}, \mathbf{h}_{-i}\right) = \sigma\left(\sum_j^P w_{ij} h_j\right), \tag{3.3}$$

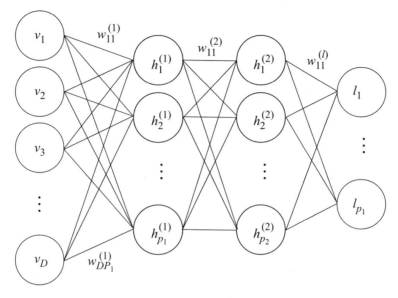

Figure 3.1: Deep Boltzmann machine with two hidden layers.

where $\sigma(x) = 1/(1 + e^{-x})$ is the logistic function. This conditional independence makes evaluation of the RBM computationally tractable. The RBM seeks to model a probability distribution over $\mathbf{v}$ using the weights W and the values of the hidden layers $\mathbf{h}$.

Exact maximum likelihood training of the RBM is not feasible, but contrastive divergence (CD) learning builds a model of the training dataset by minimizing the difference between a training sample v_i and its reconstruction $\tilde{v}_i$ [100].

The RBM concept above is extended to multiple layers by greedy layer-by-layer pretraining. The first layer weight matrix $W^{(1)}$ is trained using CD. Then, the hidden-layer $\mathbf{h}_1$ corresponding to the first layer of the network is used as training input to determine the second-layer weights $W^{(2)}$. This may be continued for an arbitrary number of hidden layers. After the pretraining, backpropagation algorithms may be used to fine-tune the network weights.

3.3   TRACKING ALGORITHMS

The problem of detecting and tracking known objects in images and video is a classic and well-studied problem in computer vision. However, adapting state-of-the-art techniques to the CS domain is non-trivial and may not be feasible for many techniques. This section focuses on tracking work which is most readily adaptable to the CS domain.

A typical visual tracking system follows a pattern in which input video frames are first preprocessed to increase image quality or enhance characteristics of the target. The image, or one or more regions of interest within the image, are then processed by a feature extraction algorithm, which converts the image into a lower-dimensional feature vector. This feature vector forms the input for the detection stage, which outputs an estimate of the confidence that a target is present at a certain location within the image. Finally, likelihood estimates across frames are fused, forming an overall estimate of the location of the target. The direct CS tracking paradigm presents problems mainly in the preprocessing and feature extraction stages, since unless reconstruction is performed, access to individual pixels is lost. Once features are extracted, tracking is able to proceed normally.

Several approaches in the literature apply sparsity or ℓ_1 based methods to the visual tracking problem. In [101], an OMP algorithm is used to extract sparse representations for visual tracking. In [102], a sparse sensing matrix is used to extract features from image regions in an approach similar to the famous Viola-Jones paradigm [103]. However, neither of these approaches operate on compressive data. Rather, they start from conventional images and apply compressive techniques for feature extraction.

3.3.1   STATE-SPACE MODEL OF THE TRACKING PROBLEM

This section introduces the system model used to describe the Kalman and particle filter algorithms. A system's time-varying state vector $\mathbf{q}_t$ is estimated from a set of measurements $\mathbf{y}_t$. $\mathbf{q}$ typically contains an estimate of target location and may also track parameters such as target speed, acceleration, scale, or pose. Optionally, the model may include a system input $\mathbf{u}_t$. In a tracking setting, this input vector could, for instance, account for known sensor platform motion. The state evolves according to the hidden Markov model (HMM),

$$\mathbf{q}_t \sim p_{\mathbf{q}_t|\mathbf{q}_{t-1},\mathbf{u}_t}\left(\mathbf{q}_t|\mathbf{q}_{t-1},\mathbf{u}_t\right) \tag{3.4}$$

$$\mathbf{y}_t \sim p_{\mathbf{y}_t|\mathbf{q}_t}\left(\mathbf{y}_t|\mathbf{q}_t\right), \tag{3.5}$$

where $p_{\mathbf{p}_i|\mathbf{p}_{i-1}}$ and $p_{\mathbf{y}_i|\mathbf{p}_i}$ are the probability density functions of the state and observed measurements, respectively. The goal of the tracking algorithm is to estimate a sequence of states $\mathbf{q}_0, \mathbf{q}_1, \ldots$ as a function of the corresponding measurements $\mathbf{y}_1, \mathbf{y}_2, \ldots$ Because of the structure of the HMM, the conditional probability of states given measurements at time t is given by the recursive relation,

$$p(\mathbf{q}_1, \ldots, \mathbf{q}_t|\mathbf{y}_1, \ldots, \mathbf{y}_t) = \frac{p(\mathbf{q}_t|\mathbf{q}_{t-1})p(\mathbf{y}_t|\mathbf{q}_t)}{p(\mathbf{y}_n|\mathbf{y}_1, \ldots, \mathbf{y}_{t-1})} p(\mathbf{q}_1, \ldots, \mathbf{q}_{t-1}|\mathbf{y}_1, \ldots, \mathbf{y}_{t-1}) \tag{3.6}$$

$$\propto p(\mathbf{q}_t|\mathbf{q}_{t-1})p(\mathbf{y}_t|\mathbf{q}_t)p(\mathbf{q}_1, \ldots, \mathbf{q}_{t-1}|\mathbf{y}_1, \ldots, \mathbf{y}_{t-1}). \tag{3.7}$$

In (3.6), $p(\cdot)$ denotes the probability density function; subscripts have been omitted for clarity and concision. This structure separates the estimate of current state $\mathbf{q}_t$ from all past measurements. Only the current measurement $\mathbf{y}$ and statistics of $\mathbf{q}_{t-1}$ are needed to calculate it.

3.3.2   KALMAN FILTER

The Kalman filter [104] gives an optimal estimate of $p(\mathbf{q}_0, \mathbf{q}_1, \ldots | \mathbf{y}_1, \mathbf{y}_2, \ldots)$ for the specific case of linear state updates and measurements in additive Gaussian noise. The general state update equation of (3.4) can be simplified to a matrix form given by

$$\mathbf{q}_t = \mathbf{F}\mathbf{q}_{t-1} + \mathbf{B}\mathbf{u}_t + \mathbf{w}_t \tag{3.8}$$

$$\mathbf{y}_t = \mathbf{H}\mathbf{q}_t + \mathbf{v}_t \tag{3.9}$$

$$\mathbf{w} \sim \mathcal{N}(\mathbf{0}, \Sigma_{\mathbf{w}}) \tag{3.10}$$

$$\mathbf{v} \sim \mathcal{N}(\mathbf{0}, \Sigma_{\mathbf{v}}). \tag{3.11}$$

Under these conditions, the *a priori* state estimate $p(\mathbf{q}_k | q_{k-1})$ and the posterior probability $\mathbf{p}(\mathbf{q}_k | q_{k-1}, \mathbf{y}_{k-1})$ are also Gaussian. The prior mean and covariance are given by

$$\hat{\mathbf{q}}_{t|t-1} = \mathbf{F}\hat{\mathbf{q}}_{t-1|t-1} + \mathbf{B}\mathbf{u}_t \tag{3.12}$$

$$\Sigma_{\mathbf{q}_{t|t-1}} = \mathbf{F}_t \Sigma_{\mathbf{q}_{t-1|t-1}} \mathbf{F}_t^T + \Sigma_{\mathbf{w}}. \tag{3.13}$$

From these quantities and the new measurement $\mathbf{y}_t$ the posterior mean and covariance are calculated. This is given in terms of the measurement residual $\tilde{\mathbf{y}}_t = \mathbf{y}_t - \mathbf{H}\mathbf{q}_{t|t-1}$ and *Kalman gain* K as

$$\hat{\mathbf{q}}_{t|t} = \hat{\mathbf{q}}_{t|t-1} + \mathbf{K}\tilde{\mathbf{y}}_t \tag{3.14}$$

$$\Sigma_{\mathbf{q}_{t|t}} = (\mathbf{I} - \mathbf{K}\mathbf{H}) \, \Sigma_{\mathbf{q}_{t|t-1}}. \tag{3.15}$$

The Kalman gain is chosen to minimize the mean square error of $\hat{\mathbf{q}}_{t|t}$; (3.15) is valid only for this optimal K. The optimal Kalman gain is given by

$$\mathbf{K} = \Sigma_{\mathbf{q}_{t|t-1}} \mathbf{H}^T \Sigma_{\tilde{\mathbf{y}}_k}^{-1}, \tag{3.16}$$

where $\Sigma_{\tilde{\mathbf{y}}_k} = \mathbf{H}\Sigma_{\mathbf{q}_{t|t-1}} \mathbf{H}^T + \Sigma_{\mathbf{v}}$ is the covariance matrix of the measurement residual $\tilde{\mathbf{y}}$.

All discussion above has assumed time-invariance of the system. The algorithm may be extended by simply replacing F, B, H, $\Sigma_{\mathbf{w}}$, and $\Sigma_{\mathbf{v}}$ by their values at time t. The Kalman filter can be used even in non-Gaussian noise. However, in this case it will not produce exact marginal probabilities.

3.3.3   PARTICLE FILTER

Particle filters are a family of algorithms for estimation of posterior probability densities. Unlike the Kalman filter and related algorithms, the particle filter is a Monte Carlo method and is able to handle arbitrary distributions. This section describes the basic vocabulary of a particle filter and details its use in visual tracking.

The particle filter estimates a time-varying state **p** given noisy measurements **y**. The state **p** evolves according to the Markov model given in (3.4).

When these PDFs are non-Gaussian or state transition equations are nonlinear, there is in general no closed-form solution for the posterior distribution of $\mathbf{p}$. However, the particle filter gives a method to sample from the distribution. q particles are generated i.i.d. according to an initial density $p_{\mathbf{p}_0}(\mathbf{p})$. At each time step, a new set of states is predicted from the previous frame, the measurement $\mathbf{y}_i$ is observed and each particle $\mathbf{p}_j$ is assigned a weight $w_{j,i}$ based on the particle's likelihood,

$$w_{j,i} = p(\mathbf{y}_i|\mathbf{x}_i)w_{j,i-1}. \tag{3.17}$$

In this way, a particle's weight grows (relative to its neighbors) during frames when its likelihood is large and shrinks during frames when its likelihood is small.

Optionally, particles may be resampled during some or all frames. This is accomplished by normalizing all the weights so that they sum to one, and forming a categorical distribution over the current particles based on these weights. A new set of particles is then drawn from this categorical distribution. Resampling is a strategy to avoid degeneracy in the particle filter, where the weights of most particles quickly approach 0 and contribute nothing to the estimation. On the other hand, when noise is small, resampling too often can lead to the estimation being dominated by one or a small number of possible states.

3.4   SUMMARY

This chapter covered several relevant problems and solutions from the image processing and computer vision literature for applications in surveillance [105, 106]. Approaches to detection and classification of images were then covered. Finally, the video tracking problem was introduced, and a selection of tracking algorithms was presented [107, 108]. These algorithms are a starting point for approaches to higher-level inference from CS images.

C H A P T E R   4

Toward Compressive Vision

In this chapter, we discuss research in the nascent field of compressive vision, namely vision algorithms which are embedded directly in the CS measurement domain and avoid the computational expense of reconstructing natural images from the measurements. This has applications for platforms that are resource-constrained, requiring a small energy budget, and a low memory and data storage footprint. While the accuracy of traditional computer vision algorithms usually is superior to those of compressive vision, reconstruction-free solutions allow the algorithm developer to trade-off accuracy for energy-efficiency or computational complexity, yielding more robust solutions for embedded vision platforms.

4.1   SMASHED FILTER

The smashed filter [12] is the compressive-domain analog of the well-known matched filter. Like the matched filter, it is the maximum likelihood classifier for the case when the observed data for each class come from a known manifold (e.g., shifted versions of a template for the case of the matched filter).

Target classification is to be performed based on compressive measurements $\mathbf{y} = \Phi(\mathbf{x} + w)$ of the spatial domain image $\mathbf{x}$. Let each element of target class $\mathcal{C}_j$ lie on a manifold defined by the function f_j and parametrized by ρ,

$$\mathbf{x} = f_j(\rho_j). \tag{4.1}$$

This leads to the following formulation of the Generalized Maximum Likelihood Classifier (GMLC),

$$\hat{j} = \underset{j}{\operatorname{argmax}}\, p\left(\mathbf{y}|\hat{\theta}_j, \mathcal{H}_i\right), \tag{4.2}$$

where $\hat{\rho}$ is the maximum likelihood estimate of the parameter vector ρ under hypothesis $\mathcal{H}_j$,

$$\hat{\rho}_j = \underset{\rho}{\operatorname{argmin}} \left\| y - \Phi f_j(\rho) \right\|_2^2. \tag{4.3}$$

Next, consider a simplistic description of the problem of visual classification. Images to be classified are shifted noisy versions of some template $\mathbf{s}_j$, and the compressive measurement y is given by

$$\mathbf{y} = \Phi(\mathbf{x} + \mathbf{w}) = \Phi(\mathbf{s}_{j,u,v} + \mathbf{w}), \tag{4.4}$$

where u and v are the x- and y-locations of the target, respectively.

When the $f_j(\theta_j)$ are shifted versions of each other as above, the smashed filter reduces to the following simple test statistic:

$$t = \mathbf{y}^T \Phi^\dagger \mathbf{s}_{j,u,v} = \mathbf{y}^T \left(\Phi\Phi^T\right)^{-1} \Phi \mathbf{s}_{j,u,v}. \tag{4.5}$$

4.1.1 PERFORMANCE OF THE SMASHED FILTER

As intuition would suggest, the correlation estimates provided by the fast smashed filter are noisy relative to those produced by the matched filter on the original image. They are also severely degraded relative to the matched filter operating on a reconstructed image. For a given sensing rate r, Signal-to-Noise (SNR) ratio will be reduced by a factor of $\sqrt{r}$ [12], i.e., a 10 dB/decade SNR penalty is imposed due to compression. In this section, we briefly compare this theoretical prediction with results from a dataset of natural images.

In order to quantify the increased noise in the smashed filter test statistic, we use the CIFAR-100 image training database, which consists of 50,000 32×32 RGB images. Each image was converted to grayscale and normalized so that its mean was zero and its ℓ_2 norm was equal to 1. Each normalized frame $\mathbf{x}_i$ was then sensed using a pseudo-random orthonormal sensing matrix Φ, generating compressive measurement $t = \mathbf{y}_i = \Phi\mathbf{x}_i$. The estimated inner product $\mathbf{y}_i^T \Phi^\dagger \mathbf{x}_i$ was then calculated, as used in the fast smashed filter test statistic. This was repeated for 10 different random drawings of Φ, and the variance over all images and all trials was calculated.

Figure 4.1 shows the effect of sensing rate r on the effective SNR of the test statistic t_{smash}, defined as

$$SNR = \frac{\mathrm{E}^2\left[t_{smash}\right]}{\mathrm{E}\left[t_{smash}^2\right]}. \tag{4.6}$$

As expected, noise variance increases as sensing rate decreases, with a 10-fold increase in sensing rate r leading to a 10 dB improvement in SNR. Note that at high sensing rates, the simulated results appear to outperform the analysis in [12]. We believe this is because the $\mathbf{x}_i$ are merely compressible and not perfectly sparse. As a result, they contain a non-sparse component which appears as noise at low sensing rates but is more accurately represented at high sensing rates. Work is underway to more accurately quantify the noise characteristics of the smashed filter for compressible signals.

4.2 SPATIO-TEMPORAL SMASHED FILTERS

In [109], the authors extended the smashed filtering framework to handle spatio-temporal data. Their framework (shown in Figure 4.2) perform action recognition by synthesizing Action MACH (Maximum Average Correlation Height) filters [95] and then smashing them into the compressed domain using the compressive matrix operator. These Action MACH filters are 3D spatio-temporal filters that are then compressed into the coded domain, and they represent a specific action from a class of actions the user wishes to detect. They called this technique

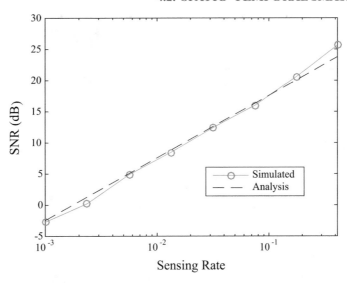

Figure 4.1: **Average SNR of smashed filter for varying sensing rate, CIFAR-100 dataset.**

spatio-temporal smashed filters (STSF). The authors provide principled ways to achieve semi-view-invariance in spatio-temporal smashed filtering, providing a way to cope with changes in viewpoint that causes misalignment in traditional MACH filtering. They do this by showing a single MACH filter in a canonical viewpoint can be transformed to any affine transform to a warped MACH filter, thus satisfying invariance under affine transformation (but not general transformation).

STFS was able to achieve a recognition rate of 54.55% at a high compression ratio of 100 on the common UCF50 dataset. On the Weizmann dataset, at a compression factor of 500, this method was able to achieve an accuracy of 78% which is much higher than a traditional method which would reconstruct the image and then perform action recognition (7.77%). In general for high compression ratios, the action recognition results shown in the paper at the time of publication were better than the state-of-the-art method applied on reconstructed images. Further, the STFS method was computationally faster than CS reconstruction and then inference, performing tasks on the Weizmann dataset in under 4 s as compared to 1500–2000 s depending on the compression factor. Thus, this was an example of a method where both accuracy and energy-efficiency was improved (at high compression factors) by using reconstruction-free compressive inference.

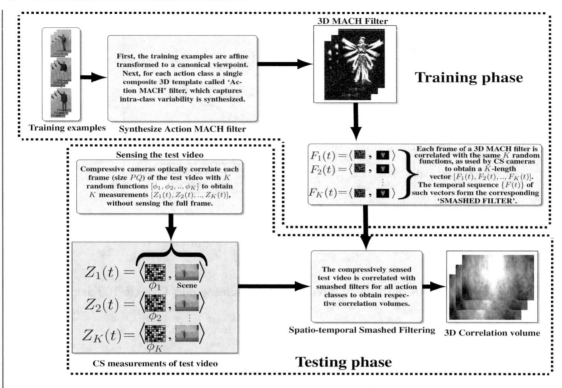

Figure 4.2: Overview of approach to action recognition from a compressively sensed test video using spatio-temporal MACH filters that are smashed into the CS measurement domain. Figure reproduced from [109].

4.3   RECONSTRUCTION-FREE COMPRESSIVE TRACKING ALGORITHM

In this section, we present a reconstruction-free compressive tracking algorithm based on combining insights from particle filtering and smashed filtering. The prototype compressive tracker is a particle filter with importance sampling updates handled by correlations with a maximum average correlation height (MACH) filter. Figure 4.3 shows a high-level block diagram showing the key components of the system and their interactions. The algorithm implements tracking-by-detection using a particle filter to maintain an estimate of both the target's detection likelihood and location, based only on data from a compressive sensing camera [110]. Correlations are estimated directly from the measured frame data $\mathbf{y}_i$ using a fast smashed filter, discussed in Section 4.1. The smashed filter produces noisy correlation estimates, reducing the effective Signal-to-noise ratio (SNR) of the detector. In order to mitigate these effects, the track-before-detect paradigm is implemented in the particle filter; the details of this approach are covered in Section 4.3.1.

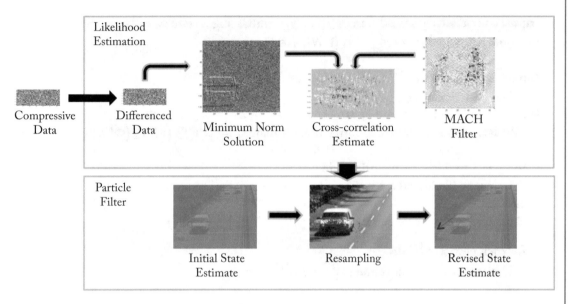

Figure 4.3: High-level block diagram of compressive tracker.

4.3.1 TRACK-BEFORE-DETECT PARTICLE FILTER

After particles are resampled, the state of each particle is updated before the next time step in accordance with the chosen Markov model of state evolution. Algorithm 4.2 gives a high-level pseudocode description of the algorithm.

State Evolution and Likelihood Model

The general operation of a particle filter is described. In this section we will discuss the state evolution and likelihood update steps used for implementation of the tracker. In this implementation, a first-order 2D motion model with scaling is used. This is defined by a linear update step with additive noise, given as,

$$\mathbf{p}_i = \mathbf{A}\mathbf{p}_{i-1} + \mathbf{w}_i. \tag{4.7}$$

The state vector $\mathbf{p}$ is given by

$$\mathbf{p} = [x\ \dot{x}\ y\ \dot{y}\ c]^T, \tag{4.8}$$

Algorithm 4.2 Particle Filter State Update Procedure.

Input: compressively sensed frames $\mathbf{y}_1, ..., \mathbf{y}_N$, initial target state $\mathbf{p}_0 = [x_0\ y_0\ \dot{x}_0\ \dot{y}_0]^T$
Output: Estimated states $\mathbf{p}_i^{(j)}$, $i \in [1, N]$, $j \in [1, q]$

Initialize particle state $\mathbf{p}_1^{(1)}...\mathbf{p}_1^{(q)} \sim p(\mathbf{p}|\mathbf{p}_0)$
Load target template $s_{0,0}$
for $i = 1$ to N {for each frame} **do**
    calculate smashed filter cross-correlation estimate $C := \mathrm{IFFT}_2\left(\mathrm{FFT}_2\left(\Phi^\dagger \mathbf{y}_i\right)\mathrm{FFT}_2^*\left(s_{0,0}\right)\right)$

    for $j = 1$ to q {for each particle} **do**
        Calculate likelihood $l_j := |C_{y_j,x_j}|$
    end for
    Resample to draw new particles $\mathbf{p}_{i+1}^{(1)}...\mathbf{p}_{i+1}^{(q)} \sim \Pr(\mathbf{p}_{i+1}^{(j)} = \mathbf{p}_i^{(k)}) = \frac{l_k}{\sum_m l_m}$

    for each particle $\mathbf{p}_i^{(j)}$ **do**
        Update state for next frame $\mathbf{p}_{i+1}^{(j)} = A\mathbf{p}_{i+1}^{(j)} + \mathbf{w}$
    end for
end for

---

where (x, y) is the target location, $(\dot{x}, \dot{y})$ is the target velocity, and c is the logarithm of the target's scale. The update matrix A is given by

$$A = \begin{bmatrix} 1 & 1 & 0 & 0 & 0 \\ 0 & 1 & 0 & 0 & 0 \\ 0 & 0 & 1 & 1 & 0 \\ 0 & 0 & 0 & 1 & 0 \\ 0 & 0 & 0 & 0 & 1 \end{bmatrix}. \tag{4.9}$$

Likelihood updates are accomplished by cross-correlation with a template s, estimated using the fast smashed filter described in Section 4.1. Let $s_{x,y}$ denote the image consisting of the template inserted at location (x, y), with zeros elsewhere. Likelihood is then modeled by

$$p(\mathbf{p}|\mathbf{y}) \propto \left| \mathbf{y}^T \Phi^\dagger s_{x,y} \right|. \tag{4.10}$$

Note that this is something of an abuse of the term "likelihood," since (4.10) does not correspond to any probability distribution on $\mathbf{y}$. This approach, however, is common in the visual tracking community.

Track-Before-Detect Implementation

Track-before-detect is a paradigm in detection wherein estimates of target state and presence/absence are maintained concurrently. This contrasts with the conventional tracking paradigm, in which a detection is declared at the same time as a target location estimate is first given. Track-before-detect may be viewed as a soft decision-based approach to this problem. In this work, we adapt the track-before-detect approach given in [111].

In the particle filter, track-before-detect is implemented as follows. State is defined using a first-order motion model as in (4.8), with the addition of a binary "alive" state a. The state vector and update model then proceeds as follows:

$$\mathbf{p} = \begin{bmatrix} \mathbf{p}_0^T & a \end{bmatrix}^T \tag{4.11}$$

$$A = \begin{bmatrix} A_0 & \mathbf{0} \\ \mathbf{0} & 1 \end{bmatrix} \tag{4.12}$$

$$\mathbf{p}_{i+1} = \begin{cases} A\mathbf{p}_i + \mathbf{w}, & a_i = 1 \\ \begin{bmatrix} \mathcal{U}(0, [w \ h]) & \mathcal{N}(0, \sigma_v^2 I)a_{i+1} \end{bmatrix}^T, & a_i = 0 \end{cases} \tag{4.13}$$

$$a_{i+1} = \begin{cases} \text{Bern}(P_b), & a_i = 0 \\ \text{Bern}(1 - P_d), & a_i = 1, \end{cases} \tag{4.14}$$

where $\mathbf{p}_0$ and A_0 are the state vector and update matrix given in (4.8) and (4.9), respectively. In other words, when $a = 1$, the particle is alive (i.e., target is present) and updates proceed as in (4.7); when $a = 0$, the particle is dead and updates sample the space of possible states. The x- and y-positions are uniformly distributed over the entire input image, and velocities are normally distributed with variance σ_v^2. a follows a Markov model in which a dead particle is "born" with probability P_b and a living particle "dies" with probability P_d.

Likelihood estimation is also defined piecewise, based on whether a particle is alive or dead. The likelihood is given by

$$p(\mathbf{y}|\mathbf{p}) = \begin{cases} \left| \mathbf{y}^T \Phi^\dagger \mathbf{s}_{x,y} \right|, & a = 1 \\ \gamma, & a = 0. \end{cases} \tag{4.15}$$

Likelihood is calculated as normal for living particles and assigned a constant value γ for dead particles. γ is selected to be less than the correlation peak height for a true detection but greater than the peak height for a false detection; changing γ allows some control over the trade-off between detection rate and false alarm rate.

4.3.2   EXPERIMENTAL RESULTS

The algorithm described in Section 4.3 was tested using simulations on several relatively simple visual datasets. Datasets were deliberately chosen for high MACH filter peak heights using a small number of filters. This allows the effect of direct Compressed sensing domain processing to be separated from the merits of the MACH/Particle filter tracking approach. Two natural video sequences from the literature were chosen for testing, the PETS2000 [112] and CDNET 2012 "Highway" [113] sequences. In Figure 4.4 both video clips show vehicles moving against a static background under relatively constant lighting conditions, allowing the use of frame differencing. The choice of vehicles as a target also minimizes the number of templates needed, since vehicles are very nearly rigid objects.

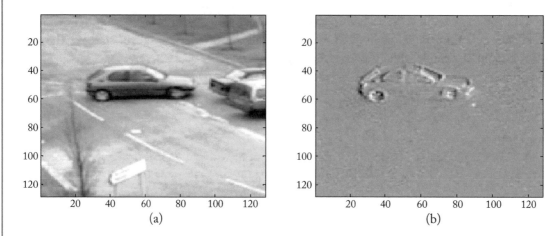

(a)   (b)

Figure 4.4: PETS2000 dataset. (a) Example frame and (b) difference frame.

The algorithm was first tested on the PETS2000 surveillance video dataset [112]. The purpose of this test was to bring the detection and tracking system to the proof of concept stage of development by achieving detection and tracking at sensing rates comparable to those required to reconstruct an image of a reasonable perceptual quality. This was accomplished, with the algorithm detecting the target at a 0.3 sensing rate.

The PETS2000 dataset was developed as a benchmark for outdoor person and vehicle tracking. In the dataset, several different cars and pedestrians cross a parking lot located in the camera's field of view. Because it includes a static background, it is a reasonable candidate for testing our algorithm. Vehicles are also seen only from a few limited angles and scales, making the detection problem more tractable. To further simplify matters, the same 201 video frames (frames 2700–2900 of the training sequence) were used for filter training and determining detector performance. We argue that reducing mismatch between the filter and the target in this way is valid, since our goal is to study the effect of the smashed filter relative to equivalent non-compressive tracker.

A single 128 × 128 pixel MACH filter was used to estimate target likelihood for the particle filter importance sampling step. The MACH filter was trained as follows: frames 2700–2900 of the differenced video sequence were cropped to a 64 × 64 pixel window containing only the target, centered as well as possible manually. These cropped target regions were then zero-padded to 128 × 128 pixels and used as input to the MACH filter generator described in (3.1). Figure 4.5

shows the spatial domain representation of the trained MACH filter, which is identifiable as the outline of a vehicle.

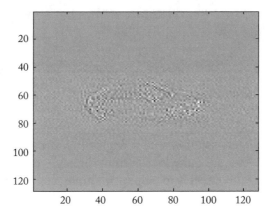

Figure 4.5: Trained MACH filter for the PETS2000 dataset.

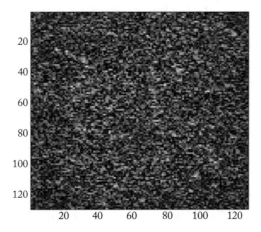

Figure 4.6: Increased noise in cross-correlation output due to smashed filter. Note that no peak from the target is visible.

Simulations were run on this dataset using the track-before-detect particle filter implementation described in Section 4.3.1.

The tracker successfully detected the target at a sensing rate of 0.3 (4915 compressive measurements for 16,384 pixels). Figure 4.8 shows the tracker's Root Mean Square Error over time for several different sensing rates. While the tracker was able to run on the PETS2000 dataset, the nature of the video sequence also served to highlight some limitations of the algorithm. Most importantly, the limitations of the MACH filter are evident, as it is only possible to

Figure 4.7: Sample frames from PETS2000 dataset.

cover a small range of the target's rotation with a single filter. Because of this, peak heights, and therefore SNR, for the detector are quite low, even in the non-compressive domain. One possible approach to mitigating this effect is to employ multiple filters, each covering a small range of target rotations. This leads to a trade-off in the number of filters, as increased performance in individual filters is balanced against increased false alarm rate across all filters.

Another limitation of the algorithm is its reliance on frame differencing. This limits the tracker to detecting only moving targets, as is evident in Figure 4.8. When the target vehicle stops and reverses direction, the tracker temporarily loses it and begins to "wander" over the image until the vehicle begins moving again. One possible approach to mitigating this effect is to maintain an estimate of the background and subtract the estimate rather than subtracting the previous frame. This approach is currently being considered.

CDNET 2012 "Highway" Sequence

The CDNET 2012 dataset [113] is a benchmark dataset for testing motion detection and tracking algorithms. It consists of several short video sequences including both indoor and outdoor scenes, but we will concentrate on a single sequence, "Highway," for tracker testing. The Highway sequence (Figure 4.9) consists of two lanes of cars approaching the camera on a highway. 1700 RGB frames, each of size 320×240 pixels, are available; pixel-level annotations are provided for frames 470–1700. This sequence was chosen because it allows the use of a single MACH filter with excellent peak height; which presents a best-case scenario for tracking with a single template. Because the video sequence contains multiple vehicle targets in most frames, no attempt was made to perform detection. Multi-target detection and tracking is an active area of research in itself and we expect multi-target tracking to be adaptable to the compressive domain with few modifications. Although detection was not formally tested, it was observed that once a target moved out of frame, the tracker quickly converged to a different target and began tracking it instead. This observation, although not conclusive, is encouraging for the potential performance of a detection algorithm on this dataset.

Two different methods, MACH and linear Support Vector Machine (SVM), were used to train a filter for cross-correlation. In both cases, frames 471–1700 of the video sequence were used for template training and frames 1–470 were used for testing. The CDNET 2012

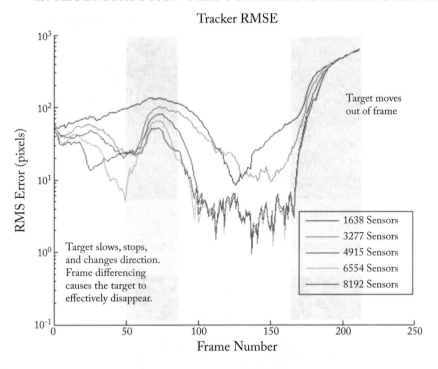

Figure 4.8: Tracker root mean square error on PETS2000 dataset.

Figure 4.9: Sample frames from CDNET 2012 highway video sequence.

dataset labels targets at the pixel level; targets were identified using blob analysis on the supplied annotation. A true example was generated by padding the target's bounding box by a factor of 1.5, cropping the resulting window, and resizing to 64×64 pixels. A false example of the same size was then generated by uniformly sampling the image windows which did not contain any pixels belonging to any target. This process is illustrated in Figure 4.11.

Process noise $\mathbf{w}$ was normally distributed with covariance matrix

$$\Sigma_{\mathbf{w}} = \mathrm{diag}\left([5\ 5\ 1\ 1\ .05]\right).\tag{4.16}$$

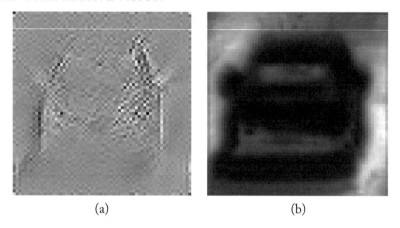

(a)   (b)

Figure 4.10: (a) MACH filter and (b) SVM filter trained for CDNET 2012 highway dataset.

(a)   (b)   (c)

Figure 4.11: CDNET 2012 highway examples. (a) Example frame, (b) pixel-level ground truth annotation, and (c) difference frame with positive and negative windows selected for training.

The CS tracker described above was compared with two other algorithms. First, a baseline non-compressive tracker was used. This approach applied the same tracking system directly to spatial domain image data, allowing virtually perfect and noise-free reconstructions relative to the compressive methods. The baseline method is thus a best-case scenario for the templates and tracking system used. Second, the reconstruct-then-track approach was examined. This method first reconstructed difference frames using the generalized approximate message passing (GAMP) algorithm [53] before performing cross-correlation with the template. Table 4.1 gives the parameters used for the GAMP reconstruction.

Figure 4.12 shows the target's ground truth path, a sample path from the tracker operating at $r = 0.1$ sensing rate, and the range of variation in 10 trials of the algorithm. Figure 4.13 shows tracker mean absolute error on the CDNET 2012 dataset as a function of sensing rate, and compares this against the baseline and GAMP reconstruction approaches. The GAMP

Table 4.1: Reconstruction parameters for GAMP algorithm

Output probability	$\mathcal{N}(y, \sigma_y^2 I)$
Output variance σ_y^2	13.01
Input probability	Bernoulli-Baussian
Input Bernoulli probability	0.005
Input Gaussian variance	500520

tracker breaks down suddenly at a sensing rate of $r = 0.01$, while the smashed filter tracker with MACH filter shows much more graceful degradation. The smashed filter tracker with SVM filter underperforms the MACH tracker. Likewise, the Hadamard sensing matrix underperforms the Gaussian case. The state-of-the-art STRUCK tracker [114] achieves an MAE of 3.6 pixels operating on non-compressive data. It slightly outperforms our algorithm, which achieves an MAE of 4.6 pixels at a sensing rate $r = 0.2$ and 6.2 pixels at $r = 0.005$.

Another metric for the performance of a tracker is the success rate. Success rate R_s, defined in [102], is the percentage of frames with overlap $o > 0.5$ between the ground truth and estimated bounding boxes. Overlap is in turn defined as the ratio of the intersection and the union between the ground truth and estimated bounding boxes:

$$o = \frac{\text{area}\left(B \cap \hat{B}\right)}{\text{area}\left(B \cup \hat{B}\right)}. \tag{4.17}$$

Success rate is chosen as a metric because it has the effect of normalizing over changes in scale while remaining simple to compute. Figure 4.14 shows the success rate of the three tracking algorithms (baseline/non-compressive, GAMP reconstruct-then-track, smashed filter) as a function of sensing rate. Again, the GAMP tracker no longer holds the target at sensing rates $r \leq 0.01$, while the smashed filter tracker continues functioning with success rate $R_s > 0.9$ at sensing rates as low as $r = 0.005$. Also, the linear SVM underperforms the MACH filter based tracker. Because STRUCK tracks a constant-sized window, the success rate metric vastly understates its performance under scale changes and is omitted from the plot.

The tracking algorithm's runtime was compared with that of the GAMP reconstruct-then-track approach. Table 4.2 summarizes the speed results for these simulations. At a sensing rate of 0.01, frame rate is increased by a factor of 10 when employing the smashed filter rather than the GAMP solver. Therefore, the smashed filter tracker clearly outperforms the GAMP tracker, both for sensing and for simulation speed.

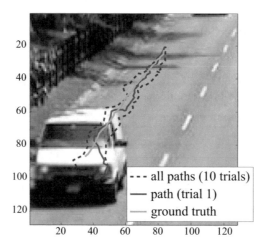

Figure 4.12: Tracker paths and ground truth for CDNET data, $r = 0.1$.

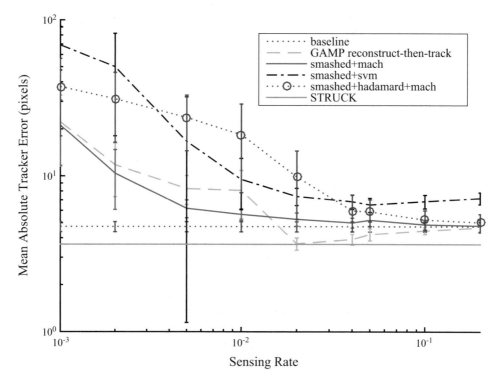

Figure 4.13: Mean absolute tracker error for highway dataset, frames 80–135.

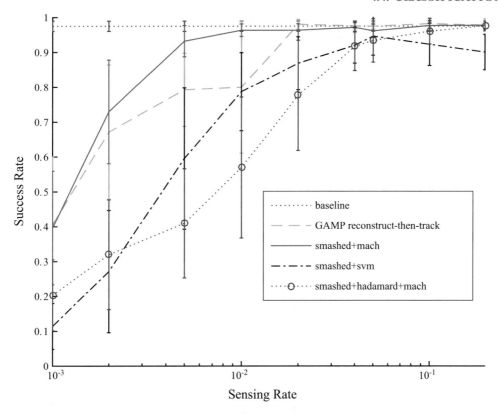

Figure 4.14: Tracker success rate for highway dataset, frames 80–135.

Table 4.2: Comparison of algorithm framerates

Algorithm	Framerate (frames/s)
Baseline	1.18
GAMP	0.112
Smashed	1.12

4.4   CLASSIFICATION

In the compressive sensing domain, research is mainly focused on reconstruction of CS images and using them to perform inference tasks. There are a handful of reconstruction-free classification methods using CS measurements. Sankaranarayanan et al. [115] use the LDS parameters obtained directly from CS measurements to perform classification. [116] and [117] use smashed correlation filters for action recognition and face recognition, respectively. In [12], author used the smashed filter to perform compressive domain classification. In [118], data-dependent "se-cant projections" were used to perform classification directly in the compressed domain. In this section, we elaborate some reconstruction free direct classification methods and their advantages on the reconstruction based methods.

4.4.1   DIRECT INFERENCE ON COMPRESSIVE MEASUREMENTS USING CNNS

Inference from compressive measurements has been a long-standing task in the field of compressive vision. A huge chunk of research has been focused to develop reconstruction algorithms to perform inference tasks. However, these algorithms have several drawbacks such as, lack of parallelization due to iterative nature, poor reconstruction at low measurement rates and being computationally expensive.

To address these drawbacks, [119] proposes a CNN-based data-driven framework to perform reconstruction free inference tasks using the discriminative nonlinear features obtained using the same framework. The method acquires compressive measurements using the single-pixel camera and transforms them into pixel space since the first layer of the architecture has a projection matrix which is the inverse of measurement matrix. The obtained pseudo image is then fed to a convolutional neural network followed by fully connected layers to perform the classification. The basic flowchart of the framework is shown in Figure 4.15.

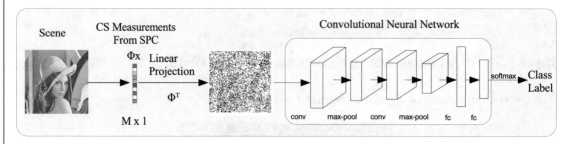

Figure 4.15: Overview of performing direct inference on compressive measurements using a CNN. The CNN block shown here is for illustration only. The specific architecture depends on the application and training data. Figure reproduced from [119].

For the MNIST hand-written digit recognition, LeNet-5 model [120] is used along with a random Gaussian measurement matrix. Backpropagation with mini-batch gradient descent is used to train the model at different measurement rates such as, 1, 0.25, 0.1, 0.05, and 0.01. While testing it is observed that at very low measurement rates the test error increase is minimal compared to the no compression case. For ILSVRC 2012 dataset [121], image recognition task, CaffeNet[2] model is used as a CNN with a low-rank column permuted Hadamard measurement matrix. The networks are trained at three different measurement rates i.e., 1, 0.25 and 0.1 and obtain the test accuracies 56.88%, 39.22%, and 29.84% respectively.

Thus, the proposed method proves that inference tasks like digit recognition and image recognition can be performed using the compressive measurements directly.

4.4.2   CLASSIFICATION ON COMPRESSIVE MEASUREMENTS VIA DEEP BOLTZMANN MACHINES

Another reconstruction free method for classification using compressive sensing measurements is developed in [122]. The DBM model is adapted for use with compressive sensing measurements. The first step is to train the DBM using non-compressive data to obtain the initial network weights W_0. For the layer-1 of DBM, weights are obtained using the following operation,

$$W_{CS}^{(1)} = \Phi W^{(1)}, \tag{4.18}$$

where Φ is a sensing matrix with orthonormal rows. These optimized weights obtained in stage 1 are further fine-tuned by backpropagation during training.

This method is tested on MNIST handwritten digit recognition task. A 2-hidden layer DBM is used for the specified dataset with 500 and 1000 hidden nodes respectively. $c = 1/r$ signifies the compression ratio where r is swept from 0.01 to 0.4. At $r = 0.4$, a confusion matrix is obtained as shown in Table 4.3. Stage-2 training using backpropagation for a chosen sensing matrix is shown in Figure 4.16. As the compression ratio was increased, there was a steady increase in error rate unlike reconstruction algorithms, which perform well up to a certain "phase transition" point and then break down.

4.5   COMPRESSIVE SENSING FOR VISUAL QUESTION ANSWERING

Visual Question Answering (VQA) consists of open-ended questions posed by a human for a given image, which requires the computer to parse the semantic information of both the question and the image to provide a written response. For example, given an image depicting a family picnic, sample questions might include "How many people are there?", "What is the color of the table?" etc. Due to the semantic analysis required to answer these questions, VQA is considered an AI complete task [123]. Contemporary VQA research has utilized deep neural-networks trained jointly on images and natural language "vectors" computed from the questions.

In [126], the authors provide an investigation into whether VQA is possible in the compressive sensing domain, i.e., with compressive measurements in place of the natural images. In Figure 4.17, the architecture for the CS-VQA network is shown. It consists of reconstruction phase followed by visual feature extraction, which is concatenated with the features extracted from question embedding and a LSTM, before going through a final network to produce the response. In Table 4.4, the results for each reconstruction method is presented.

In their results, the authors show that VQA can achieve near-equivalent performance to natural images when using advanced compressive sensing (CS) reconstruction techniques such

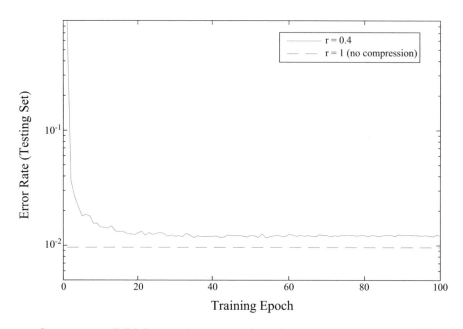

Figure 4.16: Compressive DBM error during training, for sensing rate $r = 0.4$. The CS-DBM achieves a 1.21% error rate. This is in contrast to the 0.99% error rate of the standard non-compressive DBM, shown for comparison.

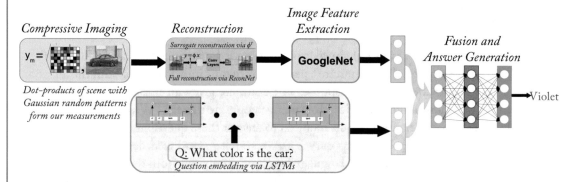

Figure 4.17: An overview of the proposed CS-VQA architecture. An image is compressively sensed and then either reconstructed using linear inversion or ReconNet [87]. The reconstructed image is used to extract visual features using GoogleNet [124]. The question is encoded using Word2vec [125] and inputted into a LSTM to form a question feature vector. The two features are concatenated and fed into a fully connected network to generate the final answer. This figure was reproduced from [126].

Table 4.3: Compressive DBM confusion matrix, $r = 0.4$

			Predicted Digit (%)							
	0	**1**	**2**	**3**	**4**	**5**	**6**	**7**	**8**	**9**
0	99.6	0	0.1	0	0	0	0.2	0.1	0	0
1	0.1	99.7	0.1	0.1	0	0	0	0	0.1	0
2	0	0.4	98.9	0	0	0	0.1	0.5	0.1	0
3	0	0	0.1	99.1	0	0.3	0	0.1	0.3	0.1
4	0	0	0.1	0	98.8	0	0.4	0	0.1	0.6
5	0	0	0	0.6	0	99.1	0.3	0	0	0
6	0.3	0.2	0	0	0.1	0.4	98.6	0	0.3	0
7	0	0.2	0.6	0	0	0	0	98.9	0	0.3
8	0.3	0	0.2	0.2	0.1	0.2	0.1	0.2	98.4	0.3
9	0.2	0	0	0	0.6	0.3	0	0.5	0.4	98.0

(Row labels along the left: **True Digit**)

Table 4.4: Open-ended VQA v2.0 results with various CS reconstructions, and their corresponding accuracy (%). Table reproduced from [126].

CS Reconstruction Method	Question Type			
	All	**Yes/No**	**Number**	**Other**
Linear Inversion	48.92	70.61	33.13	36.58
ReconNet	49.85	70.50	33.32	38.52
Oracle VQA [127]				
LSTM + VGG	54.22	73.46	35.18	41.83
Question Only	44.26	67.01	31.55	27.37

as ReconNet with a performance gap of only 5% points at measurement rate MR = 0.25. By performing direct inference, they report faster processing times over approaches that require reconstruction and smaller networks. This work opens a new avenue for compressive vision where multimodal tasks across domains can be explored for compresive measurements.

CHAPTER 5

Conclusion

In this book, we have introduced new research directions in performing inference on compressively sensed data. Our aim has been two-fold: to situate the work within the context of compressive sensing and computer vision, as well as showing new research papers pushing the boundaries of what is possible in the field. Compressive vision has two main goals: first to improve the accuracy of inference on compressive measurements, and second to reduce the computational burden of making them. These algorithmic results can be ported to several compressive sensing hardware platforms from the single-pixel camera to infrared, multispectral, and magnetic resonance images. Several tools were surveyed as useful for this task including smashed filters, deep Boltzmann machines, and particle filter tracking.

In Chapter 2, we presented the basics of compressive sensing and in Chapter 3 we discussed computer vision and image processing applications. In Chapter 4, we discuss the use of smashed filters and particle filters in compressive sensing applications. We also discuss two motivating examples of compressive inference we presented earlier in the book. In Section 4.3, we describe an automated vehicle tracking algorithm which follows targets in CS video at computational speeds comparable to an equivalent non-CS algorithm. This algorithm successfully holds targets at sensing rates below that required for successful single-frame image reconstruction. The tracker operates in the tracking by detection paradigm and employs a particle filter to maintain and update estimates of target state. In each frame, detection is performed with the fast smashed filter, which generates noisy estimates of image correlation with a target template. By performing background subtraction and training templates based on difference frames, clutter was dramatically reduced, at the cost of limiting the algorithm's application to moving targets with stationary backgrounds. In Section 4.4.2 considers deep Boltzmann machines as one possible way to translate sophisticated machine learning algorithms to the reconstruction-free CS classification and detection problem. A DBM model was trained on non-compressive data, and its network weights were projected to an equivalent CS model. This equivalent model was then fine-tuned by further training on compressively sensed versions of the same training data, generating a CS-DBM network trained specifically for a single sensing matrix. This fine-tuning step was found to greatly reduce classification error. Both these examples show the rich area for exploration which reconstruction-free compressive vision can play.

5.1   FINAL REMARKS AND FURTHER READING

This work presents in part an attempt to broaden the applications of compressed sensing, particularly in areas where computational power is limited such as reconstruction-free compressive vision. Applications in video-based surveillance and autonomous navigation were targeted [128]. By developing algorithms that perform inference on CS data at computational costs comparable to computer vision algorithms operating on non-CS, new applications of CS sensors become feasible. Our hope is that this work leads to the widespread adoption of cost-effective high-performance compressed sensing systems in a diverse range of imaging applications.

The field of compressive vision and deep learning is vast and we encourage the reader to go through the extensive bibliography provided to expose themselves to different methods in these areas. For further information regarding signal compression applications, readers can refer to [129–133]. Deep learning has gained a huge impetus in several applications in past decade including shading prediction, dictionary learning and surveillance [134–139]. Deep learning algorithms have found huge success in image classification, object detection, time series prediction and natural language processing [140–143]. For further reading about various architectures of deep learning, readers can refer to [144–147].

Bibliography

[1] Dharmpal Takhar, Jason N. Laska, Michael B. Wakin, Marco F. Duarte, Dror Baron, Shriram Sarvotham, Kevin F. Kelly, and Richard G. Baraniuk, A new compressive imaging camera architecture using optical-domain compression, in *Computational Imaging IV*. International Society for Optics and Photonics, vol. 6065, no. 606509, 2006. DOI: 10.1117/12.659602 1, 9

[2] Jayaraman J. Thiagarajan and Andreas Spanias, Learning dictionaries for local sparse coding in image classification, in *45th IEEE Asilomar Conference on Signals, Systems and Computers (ASILOMAR)*, pp. 2014–2018, Pacific Grove, CA, November 2011. DOI: 10.1109/acssc.2011.6190379 1

[3] David L. Donoho, Compressed sensing, in *IEEE Transactions on Information Theory*, vol. 52, no. 4, pp. 1289–1306, 2006. DOI: 10.1109/tit.2006.871582 1

[4] Marco F. Duarte, Mark A. Davenport, Dharmpal Takhar, Jason N. Laska, Ting Sun, Kevin F. Kelly, and Richard G. Baraniuk, Single-pixel imaging via compressive sampling, in *IEEE Signal Processing Magazine*, vol. 25, no. 2, pp. 83–91, 2008. DOI: 10.1109/msp.2007.914730 1

[5] Ingrid Daubechies, The wavelet transform, time-frequency localization and signal analysis, in *IEEE Transactions on Information Theory*, vol. 36, no. 5, pp. 961–1005, 1990. DOI: 10.1515/9781400827268.442 1

[6] Bruce D. Lucas and Takeo Kanade, An iterative image registration technique with an application to stereo vision (darpa), in *Proc. of the DARPA Image Understanding Workshop*, pp. 121–130, April 1981. 2

[7] Bruno A. Olshausen and David J. Field, Sparse coding with an overcomplete basis set: A strategy employed by v1?, in *Vision Research*, vol. 37, no. 23, pp. 3311–3325, 1997. DOI: 10.1016/s0042-6989(97)00169-7 6

[8] Enrique G. Ortiz and Brian C. Becker, Face recognition for web-scale datasets, in *Computer Vision and Image Understanding*, vol. 118, pp. 153–170, 2014. DOI: 10.1016/j.cviu.2013.09.004 6

[9] Emmanuel J. Candes, The restricted isometry property and its implications for compressed sensing, in *Comptes Rendus Mathematique*, vol. 346, no. 9–10, pp. 589–592, 2008. DOI: 10.1016/j.crma.2008.03.014 6, 28, 30

[10] Emmanuel J. Candes and Terence Tao, Near-optimal signal recovery from random projections: Universal encoding strategies?, in *IEEE Transactions on Information Theory*, vol. 52, no. 12, pp. 5406–5425, 2006. DOI: 10.1109/tit.2006.885507 6

[11] Lu Gan, Cong Ling, Thong T. Do, and Trac D. Tran, Analysis of the statistical restricted isometry property for deterministic sensing matrices using stein's method, in *Preprint*, vol. 190, 2009. 6

[12] Mark A. Davenport, Marco F. Duarte, Michael B. Wakin, Jason N. Laska, Dharmpal Takhar, Kevin F. Kelly, and Richard G. Baraniuk, The smashed filter for compressive classification and target recognition, in *Computational Imaging V*. International Society for Optics and Photonics, vol. 6498, no. 64980H, 2007. DOI: 10.1117/12.714460 6, 45, 46, 59

[13] Robert Calderbank, Stephen Howard, and Sina Jafarpour, Construction of a large class of deterministic sensing matrices that satisfy a statistical isometry property, in *IEEE Journal of Selected Topics in Signal Processing*, vol. 4, no. 2, pp. 358–374, 2010. DOI: 10.1109/jstsp.2010.2043161 6

[14] Gregory K. Wallace, The JPEG still picture compression standard, in *IEEE Transactions on Consumer Electronics*, vol. 38, no. 1, pp. XVIII–XXXIV, 1992. DOI: 10.1145/103085.103089 6

[15] Athanassios Skodras, Charilaos Christopoulos, and Touradj Ebrahimi, The JPEG 2000 still image compression standard, in *IEEE Signal Processing Magazine*, vol. 18, no. 5, pp. 36–58, 2001. DOI: 10.1109/79.952804 6

[16] Richard G. Baraniuk, Volkan Cevher, Marco F. Duarte, and Chinmay Hegde, Model-based compressive sensing, in *IEEE Transactions on Information Theory*, vol. 56, no. 4, pp. 1982–2001, 2010. DOI: 10.1109/tit.2010.2040894 6, 8, 32

[17] Shihao Ji, Ya Xue, and Lawrence Carin, Bayesian compressive sensing, in *IEEE Transactions on Signal Processing*, vol. 56, no. 6, pp. 2346, 2008. DOI: 10.1109/tsp.2007.914345 7

[18] Mário Figueiredo, Adaptive sparseness using Jeffreys prior, in *Advances in Neural Information Processing Systems*, pp. 697–704, 2002. 8

[19] Sivan Gleichman and Yonina C. Eldar, Blind compressed sensing, in *IEEE Transactions on Information Theory*, vol. 57, no. 10, pp. 6958–6975, 2011. DOI: 10.1109/tit.2011.2165821 9

[20] Michael B. Wakin, Jason N. Laska, Marco F. Duarte, Dror Baron, Shriram Sarvotham, Dharmpal Takhar, Kevin F. Kelly, and Richard G. Baraniuk, An architecture for compressive imaging, in *IEEE International Conference on Image Processing*, pp. 1273–1276, 2006. DOI: 10.1109/icip.2006.312577 9

[21] Matthew A. Herman, Donna E. Hewitt, Tyler H. Weston, and Lenore McMackin, Overlap patterns and image stitching for multiple-detector compressive-sensing camera, U.S. Patent 8,970,740, March 3, 2015. 10

[22] Huaijin Chen, M. Salman Asif, Aswin C. Sankaranarayanan, and Ashok Veeraraghavan, FPA-CS: Focal plane array-based compressive imaging in short-wave infrared, in *Proc. of the IEEE Conference on Computer Vision and Pattern Recognition*, pp. 2358–2366, 2015. DOI: 10.1109/cvpr.2015.7298849 10

[23] M. E. Gehm, R. John, D. J. Brady, R. M. Willett, and T. J. Schulz, Single-shot compressive spectral imaging with a dual-disperser architecture, in *Optics Express*, vol. 15, no. 21, pp. 14013–14027, 2007. DOI: 10.1364/oe.15.014013 10

[24] Ashwin A. Wagadarikar, Nikos P. Pitsianis, Xiaobai Sun, and David J. Brady, Video rate spectral imaging using a coded aperture snapshot spectral imager, in *Optics Express*, vol. 17, no. 8, pp. 6368–6388, April 2009. DOI: 10.1364/oe.17.006368 10

[25] Ge Wang, Yoram Bresler, and Vasilis Ntziachristos, Guest editorial compressive sensing for biomedical imaging, in *IEEE Transactions on Medical Imaging*, vol. 30, no. 5, pp. 1013–1016, 2011. DOI: 10.1109/tmi.2011.2145070 10

[26] Oren N. Jaspan, Roman Fleysher, and Michael L. Lipton, Compressed sensing MRI: A review of the clinical literature, in *The British Journal of Radiology*, vol. 88, no. 1056, pp. 20150487, 2015. DOI: 10.1259/bjr.20150487 10

[27] Jayaraman J. Thiagarajan, Karthikeyan N. Ramamurthy, Berkay Kanberoglu, David Frakes, Kevin Bennett, and Andreas Spanias, Measuring glomerular number from kidney MRI images, in *SPIE Medical Imaging*, 2016. DOI: 10.1117/12.2216753 11

[28] Jayaraman J. Thiagarajan, Karthikeyan N. Ramamurthy, David Frakes, and Andreas Spanias, An algorithm to estimate glomerular number from kidney magnetic resonance images, U.S. Patent 9,779,497, October 3, 2017. 11

[29] Gary H. Glover, Simple analytic spiral k-space algorithm, in *Magnetic Resonance in Medicine: An Official Journal of the International Society for Magnetic Resonance in Medicine*, vol. 42, no. 2, pp. 412–415, 1999. DOI: 10.1002/(sici)1522-2594(199908)42:2<412::aid-mrm25>3.0.co;2-u 11

[30] Michael Lustig, David L. Donoho, Juan M. Santos, and John M. Pauly, Compressed Sensing MRI, in *IEEE Signal Processing Magazine*, vol. 25, no. 2, pp. 72–82, 2008. DOI: 10.1109/msp.2007.914728 11

[31] Michael Lustig, David Donoho, and John M. Pauly, Sparse MRI: The application of compressed sensing for rapid MR imaging, in *Magnetic Resonance in Medicine: An Official Journal of the International Society for Magnetic Resonance in Medicine*, vol. 58, no. 6, pp. 1182–1195, 2007. DOI: 10.1002/mrm.21391 11

[32] Ti-Chiun Chang, Lin He, and Tong Fang, MR image reconstruction from sparse radial samples using Bregman iteration, in *Proc. of the 13th Annual Meeting of ISMRM, Seattle*, vol. 696, pp. 482, 2006. 11

[33] Jong Chul Ye, Sungho Tak, Yeji Han, and Hyun Wook Park, Projection reconstruction MR imaging using focuss, in *Magnetic Resonance in Medicine: An Official Journal of the International Society for Magnetic Resonance in Medicine*, vol. 57, no. 4, pp. 764–775, 2007. DOI: 10.1002/mrm.21202

[34] Kai Tobias Block, Martin Uecker, and Jens Frahm, Undersampled radial MRI with multiple coils. Iterative image reconstruction using a total variation constraint, in *Magnetic Resonance in Medicine: An Official Journal of the International Society for Magnetic Resonance in Medicine*, vol. 57, no. 6, pp. 1086–1098, 2007. DOI: 10.1002/mrm.21236

[35] Sean B. Fain, Walter Block, Andrew Barger, and Charles A. Mistretta, Correction for artifacts in 3D angularly undersampled MR projection reconstruction, in *9th Annual Meeting of ISMRM, Glasgow*, p. 759, Citeseer, 2001. 11

[36] Juan M. Santos, Charles H. Cunningham, Michael Lustig, Brian A. Hargreaves, Bob S. Hu, Dwight G. Nishimura, and John M. Pauly, Single breath-hold whole-heart MRA using variable-density spirals at 3T, in *Magnetic Resonance in Medicine: An Official Journal of the International Society for Magnetic Resonance in Medicine*, vol. 55, no. 2, pp. 371–379, 2006. DOI: 10.1002/mrm.20765 11

[37] Michael Lustig, Jin Hyung Lee, David L. Donoho, and John M. Pauly, Faster imaging with randomly perturbed, under-sampled spirals and l1 reconstruction, in *Proc. of the 13th Annual Meeting of ISMRM*, p. 685, Citeseer, Miami Beach, FL, 2005. 11

[38] S. S. Vasanawala, M. J. Murphy, Marcus T. Alley, P. Lai, Kurt Keutzer, John M. Pauly, and Michael Lustig, Practical parallel imaging compressed sensing MRI: Summary of two years of experience in accelerating body MRI of pediatric patients, in *IEEE International Symposium on Biomedical Imaging: From Nano to Macro*, pp. 1039–1043, 2011. DOI: 10.1109/isbi.2011.5872579 11

[39] Benjamin P. Berman, Abhishek Pandey, Zhitao Li, Lindsie Jeffries, Theodore P. Trouard, Isabel Oliva, Felipe Cortopassi, Diego R. Martin, Maria I. Altbach, and Ali Bilgin, Volumetric MRI of the lungs during forced expiration, in *Magnetic Resonance in Medicine*, vol. 75, no. 6, pp. 2295–2302, 2016. DOI: 10.1002/mrm.25798 11

[40] Kshitij Marwah, Gordon Wetzstein, Yosuke Bando, and Ramesh Raskar, Compressive light field photography using overcomplete dictionaries and optimized projections, in *ACM Transactions on Graphics (TOG)*, vol. 32, no. 4, p. 46, 2013. DOI: 10.1145/2461912.2461914 11

[41] Matthew Hirsch, Sriram Sivaramakrishnan, Suren Jayasuriya, Albert Wang, Alyosha Molnar, Ramesh Raskar, and Gordon Wetzstein, A switchable light field camera architecture with angle sensitive pixels and dictionary-based sparse coding, in *IEEE International Conference on Computational Photography (ICCP)*, pp. 1–10, 2014. DOI: 10.1109/iccphot.2014.6831813 11

[42] David J. Brady, Kerkil Choi, Daniel L. Marks, Ryoichi Horisaki, and Sehoon Lim, Compressive holography, in *Optics Express*, vol. 17, no. 15, pp. 13040–13049, 2009. DOI: 10.1364/oe.17.013040 12

[43] Zihao Wang, Leonidas Spinoulas, Kuan He, Lei Tian, Oliver Cossairt, Aggelos K. Katsaggelos, and Huaijin Chen, Compressive holographic video, in *Optics Express*, vol. 25, no. 1, pp. 250–262, January 2017. DOI: 10.1364/oe.25.000250 12

[44] Lei Tian, Xiao Li, Kannan Ramchandran, and Laura Waller, Multiplexed coded illumination for fourier ptychography with an led array microscope, in *Biomedical Optics Express*, vol. 5, no. 7, pp. 2376–2389, 2014. DOI: 10.1364/boe.5.002376 12

[45] Mário A. T. Figueiredo, Robert D. Nowak, and Stephen J. Wright, Gradient projection for sparse reconstruction: Application to compressed sensing and other inverse problems, in *IEEE Journal of Selected Topics in Signal Processing*, vol. 1, no. 4, pp. 586–597, 2007. DOI: 10.1109/jstsp.2007.910281 12

[46] Ewout van den Berg and Michael P. Friedlander, SPGL1: A solver for large-scale sparse reconstruction, 2007. 12, 26

[47] Yagyensh Chandra Pati, Ramin Rezaiifar, and Perinkulam Sambamurthy Krishnaprasad, Orthogonal matching pursuit: Recursive function approximation with applications to wavelet decomposition, in *IEEE 27th Asilomar Conference on Signals, Systems and Computers*, pp. 40–44, 1993. DOI: 10.1109/acssc.1993.342465 12, 26

[48] Joel A. Tropp and Anna C. Gilbert, Signal recovery from random measurements via orthogonal matching pursuit, in *IEEE Transactions on Information Theory*, vol. 53, no. 12, pp. 4655–4666, 2007. DOI: 10.1109/tit.2007.909108 13

[49] Deanna Needell and Roman Vershynin, Uniform uncertainty principle and signal recovery via regularized orthogonal matching pursuit, in *Foundations of Computational Mathematics*, vol. 9, no. 3, pp. 317–334, 2009. DOI: 10.1007/s10208-008-9031-3

[50] David L. Donoho, Yaakov Tsaig, Iddo Drori, and Jean-Luc Starck, Sparse solution of underdetermined systems of linear equations by stagewise orthogonal matching pursuit, in *IEEE Transactions on Information Theory*, vol. 58, no. 2, pp. 1094–1121, 2012. DOI: 10.1109/tit.2011.2173241 12

[51] Deanna Needell and Joel A. Tropp, CoSaMP: Iterative signal recovery from incomplete and inaccurate samples, in *Applied and Computational Harmonic Analysis*, vol. 26, no. 3, pp. 301–321, 2009. DOI: 10.1145/1859204.1859229 13, 21, 26, 28

[52] Kiryung Lee and Yoram Bresler, ADMiRA: Atomic decomposition for minimum rank approximation, in *IEEE Transactions on Information Theory*, vol. 56, no. 9, pp. 4402–4416, 2010. DOI: 10.1109/tit.2010.2054251 13

[53] Sundeep Rangan, Generalized approximate message passing for estimation with random linear mixing, in *IEEE International Symposium on Information Theory Proceedings (ISIT)*, pp. 2168–2172, 2011. DOI: 10.1109/isit.2011.6033942 13, 18, 19, 26, 56

[54] Gregory F. Cooper, The computational complexity of probabilistic inference using Bayesian belief networks, in *Artificial Intelligence*, vol. 42, no. 2–3, pp. 393–405, 1990. DOI: 10.1016/0004-3702(90)90060-d 14

[55] Judea Pearl, *Reverend Bayes on Inference Engines: A Distributed Hierarchical Approach*, Cognitive Systems Laboratory, School of Engineering and Applied Science, University of California, Los Angeles, CA, 1982. 14

[56] Yair Weiss, Correctness of local probability propagation in graphical models with loops, in *Neural Computation*, vol. 12, no. 1, pp. 1–41, 2000. DOI: 10.1162/089976600300015880 16, 17

[57] Robert J. McEliece, David J. C. MacKay, and Jung-Fu Cheng, Turbo decoding as an instance of pearl's "belief propagation" algorithm, in *IEEE Journal on Selected Areas in Communications*, vol. 16, no. 2, pp. 140–152, 1998. DOI: 10.1109/49.661103 16

[58] Kevin P. Murphy, Yair Weiss, and Michael I. Jordan, Loopy belief propagation for approximate inference: An empirical study, in *Proc. of the 15th Conference on Uncertainty in Artificial Intelligence*, pp. 467–475, Morgan Kaufmann Publishers Inc., 1999. 16

[59] Andrea Montanari, Balaji Prabhakar, and David Tse, Belief propagation based multi-user detection, in *ArXiv Preprint cs/0510044*, 2005. 17

[60] Sundeep Rangan, Estimation with random linear mixing, belief propagation and compressed sensing, in *44th Annual IEEE Conference on Information Sciences and Systems (CISS)*, pp. 1–6, 2010. DOI: 10.1109/ciss.2010.5464768 17

[61] Dongning Guo and Chih-Chun Wang, Asymptotic mean-square optimality of belief propagation for sparse linear systems, in *IEEE Information Theory Workshop (ITW)*, pp. 194–198, 2006. DOI: 10.1109/itw2.2006.323786 17

[62] David L. Donoho, Arian Maleki, and Andrea Montanari, Message-passing algorithms for compressed sensing, in *Proc. of the National Academy of Sciences*, vol. 106, no. 45, pp. 18914–18919, 2009. DOI: 10.1109/itwksps.2010.5503228 17, 18

[63] David L. Donoho, Arian Maleki, and Andrea Montanari, Message passing algorithms for compressed sensing: I. motivation and construction, in *IEEE Information Theory Workshop on Information Theory (ITW)*, pp. 1–5, 2010. DOI: 10.1109/itwksps.2010.5503193 17

[64] Jeremy P. Vila and Philip Schniter, Expectation-maximization Gaussian-mixture approximate message passing, in *IEEE Transactions on Signal Processing*, vol. 61, no. 19, pp. 4658–4672, 2013. DOI: 10.1109/ciss.2012.6310932 20, 26

[65] Jason T. Parker, Philip Schniter, and Volkan Cevher, Bilinear generalized approximate message passing—part I: Derivation, in *IEEE Transactions on Signal Processing*, vol. 62, no. 22, pp. 5839–5853, 2014. DOI: 10.1109/tsp.2014.2357776 20

[66] Andrew E. Waters, Aswin C. Sankaranarayanan, and Richard Baraniuk, SpaRCS: Recovering low-rank and sparse matrices from compressive measurements, in *Advances in Neural Information Processing Systems*, pp. 1089–1097, 2011. 21

[67] Aswin C. Sankaranarayanan, Christoph Studer, and Richard G. Baraniuk, CS-MUVI: Video compressive sensing for spatial-multiplexing cameras, in *IEEE International Conference on Computational Photography (ICCP)*, pp. 1–10, 2012. DOI: 10.1109/iccphot.2012.6215212 22

[68] Volkan Cevher, Aswin Sankaranarayanan, Marco F. Duarte, Dikpal Reddy, Richard G. Baraniuk, and Rama Chellappa, Compressive sensing for background subtraction, in *European Conference on Computer Vision*, pp. 155–168, Springer, 2008. DOI: 10.1007/978-3-540-88688-4_12 22

[69] Nathan Jacobs, Stephen Schuh, and Robert Pless, Compressive sensing and differential image-motion estimation, in *IEEE International Conference on Acoustics Speech and Signal Processing (ICASSP)*, pp. 718–721, 2010. DOI: 10.1109/icassp.2010.5495053 22

[70] Vijayaraghavan Thirumalai and Pascal Frossard, Correlation estimation from compressed images, in *Journal of Visual Communication and Image Representation*, vol. 24, no. 6, pp. 649–660, 2013. DOI: 10.1016/j.jvcir.2011.12.004 22

[71] Xiaohui Shen and Ying Wu, Sparsity model for robust optical flow estimation at motion discontinuities, 2010. DOI: 10.1109/cvpr.2010.5539944 22

[72] Petr Tichavsky, Carlos H. Muravchik, and Arye Nehorai, Posterior Cramér-Rao bounds for discrete-time nonlinear filtering, in *IEEE Transactions on Signal Processing*, vol. 46, no. 5, pp. 1386–1396, 1998. DOI: 10.1109/78.668800 23

[73] Hadi Zayyani, Massoud Babaie-Zadeh, and Christian Jutten, Bayesian Cramér-Rao bound for noisy non-blind and blind compressed sensing, in *ArXiv Preprint ArXiv:1005.4316*, 2010. 25

[74] David Donoho and Jared Tanner, Counting faces of randomly projected polytopes when the projection radically lowers dimension, in *Journal of the American Mathematical Society*, vol. 22, no. 1, pp. 1–53, 2009. DOI: 10.1090/s0894-0347-08-00600-0 26

[75] David Donoho and Jared Tanner, Observed universality of phase transitions in high-dimensional geometry, with implications for modern data analysis and signal processing, in *Philosophical Transactions of the Royal Society of London A: Mathematical, Physical and Engineering Sciences*, vol. 367, no. 1906, pp. 4273–4293, 2009. DOI: 10.1098/rsta.2009.0152 26

[76] Khanh Do Ba, Piotr Indyk, Eric Price, and David P. Woodruff, Lower bounds for sparse recovery, in *Proc. of the 21st Annual ACM-SIAM Symposium on Discrete Algorithms*, pp. 1190–1197, 2010. DOI: 10.1137/1.9781611973075.95 26

[77] Emmanuel Candes and Justin Romberg, l1-magic: Recovery of sparse signals via convex programming, vol. 4, pp. 14, 2005. www.acm.caltech.edu/l1magic/downloads/l1magic.pdf 26

[78] David L. Donoho, Arian Maleki, and Andrea Montanari, The noise-sensitivity phase transition in compressed sensing, in *IEEE Transactions on Information Theory*, vol. 57, no. 10, pp. 6920–6941, 2011. DOI: 10.1109/tit.2011.2165823 27, 30

[79] Eric W. Tramel, Angélique Drémeau, and Florent Krzakala, Approximate message passing with restricted Boltzmann machine priors, in *Journal of Statistical Mechanics: Theory and Experiment*, no. 7, pp. 073401, 2016. DOI: 10.1088/1742-5468/2016/07/073401 32

[80] Eric W. Tramel, Andre Manoel, Francesco Caltagirone, Marylou Gabrié, and Florent Krzakala, Inferring sparsity: Compressed sensing using generalized restricted Boltzmann machines, in *IEEE Information Theory Workshop (ITW)*, pp. 265–269, 2016. DOI: 10.1109/itw.2016.7606837 32

[81] Rick Chartrand and Wotao Yin, Iterative reweighted algorithms for compressive sensing, *Technical Report*, 2008. DOI: 10.1109/icassp.2008.4518498 32

[82] Ali Mousavi, Ankit B. Patel, and Richard G. Baraniuk, A deep learning approach to structured signal recovery, in *53rd Annual Allerton IEEE Conference on Communication, Control, and Computing (Allerton)*, pp. 1336–1343, 2015. DOI: 10.1109/allerton.2015.7447163 33

[83] Ali Mousavi and Richard G. Baraniuk, Learning to invert: Signal recovery via deep convolutional networks, in *IEEE International Conference on Acoustics, Speech and Signal Processing (ICASSP)*, pp. 2272–2276, 2017. DOI: 10.1109/icassp.2017.7952561 33, 34

[84] Ali Mousavi, Gautam Dasarathy, and Richard G. Baraniuk, DeepCodec: Adaptive sensing and recovery via deep convolutional neural networks, in *ArXiv Preprint ArXiv:1707.03386*, 2017. DOI: 10.1109/allerton.2017.8262812 33, 34

[85] Akshat Dave, Anil Kumar, and Kaushik Mitra, Compressive image recovery using recurrent generative model, in *IEEE International Conference on Image Processing (ICIP)*, pp. 1702–1706, 2017. DOI: 10.1109/icip.2017.8296572 34

[86] Akshat Dave, Anil Kumar Vadathya, Ramana Subramanyam, Rahul Baburajan, and Kaushik Mitra, Solving inverse computational imaging problems using deep pixel-level prior, in *ArXiv Preprint ArXiv:1802.09850*, 2018. DOI: 10.1109/tci.2018.2882698 35

[87] Suhas Lohit, Kuldeep Kulkarni, Ronan Kerviche, Pavan Turaga, and Amit Ashok, Convolutional neural networks for noniterative reconstruction of compressively sensed images, in *IEEE Transactions on Computational Imaging*, vol. 4, no. 3, pp. 326–340, 2018. DOI: 10.1109/tci.2018.2846413 36, 62

[88] Abhijit Mahalanobis and Robert Muise, Object specific image reconstruction using a compressive sensing architecture for application in surveillance systems, in *IEEE Transactions on Aerospace and Electronic Systems*, vol. 45, no. 3, 2009. DOI: 10.1109/taes.2009.5259191 37

[89] David A. Forsyth and Jean Ponce, *Computer Vision: A Modern Approach*, Prentice Hall Professional Technical Reference, 2002. 39

[90] Alan C. Bovik, *Handbook of Image and Video Processing*, Academic Press, 2010. 39

[91] Pawan Sinha, Benjamin Balas, Yuri Ostrovsky, and Richard Russell, Face recognition by humans: Nineteen results all computer vision researchers should know about, in *Proc. of the IEEE*, vol. 94, no. 11, pp. 1948–1962, 2006. 39
DOI: 10.1109/jproc.2006.884093

[92] Pavan Turaga, Rama Chellappa, Venkatramana S. Subrahmanian, and Octavian Udrea, Machine recognition of human activities: A survey, in *IEEE Transactions on Circuits and Systems for Video Technology*, vol. 18, no. 11, pp. 1473, 2008. DOI: 10.1109/tcsvt.2008.2005594 39

[93] Benjamin Coifman, David Beymer, Philip McLauchlan, and Jitendra Malik, A real-time computer vision system for vehicle tracking and traffic surveillance, in *Transportation Research Part C: Emerging Technologies*, vol. 6, no. 4, pp. 271–288, 1998. DOI: 10.1016/s0968-090x(98)00019-9 39

[94] Hanying Zhou, Casey Hughlett, Jay C. Hanan, Thomas Lu, and Tien-Hsin Chao, Development of streamlined OT-MACH-based ATR algorithm for grayscale optical correlator, in *Optical Pattern Recognition XVI*, vol. 5816, pp. 78–84, International Society for Optics and Photonics, 2005. DOI: 10.1117/12.607969 40

[95] Mikel D. Rodriguez, Javed Ahmed, and Mubarak Shah, Action MACH a spatio-temporal maximum average correlation height filter for action recognition, in *IEEE Conference on Computer Vision and Pattern Recognition, (CVPR)*, pp. 1–8, 2008. DOI: 10.1109/cvpr.2008.4587727 40, 46

[96] Abhijit Mahalanobis and Bhagavatula Vijaya Kumar, Optimality of the maximum average correlation height filter for detection of targets in noise, in *Optical Engineering*, vol. 36, no. 10, pp. 2642–2649, 1997. DOI: 10.1117/1.601314 40

[97] Oliver C. Johnson, Weston Edens, Thomas T. Lu, and Tien-Hsin Chao, Optimization of OT-MACH filter generation for target recognition, in *Optical Pattern Recognition XX*, vol. 7340, pp. 734008, International Society for Optics and Photonics, 2009. DOI: 10.1117/12.820950 40

[98] Abhijit Mahalanobis, B. V. K. Vijaya Kumar, Sewoong Song, S. R. F. Sims, and J. F. Epperson, Unconstrained correlation filters, in *Applied Optics*, vol. 33, no. 17, pp. 3751–3759, 1994. DOI: 10.1364/ao.33.003751 40

[99] Ruslan Salakhutdinov and Geoffrey Hinton, Deep Boltzmann machines, in *Proc. of the 12th International Conference on Artificial Intelligence and Statistics*, pp. 16–18 April, vol. 5 of *Proc. of Machine Learning Research*, pp. 448–455, 2009. 40

[100] Geoffrey E. Hinton, Training products of experts by minimizing contrastive divergence, in *Neural Computation*, vol. 14, no. 8, pp. 1771–1800, 2002. DOI: 10.1162/089976602760128018 41

[101] Hanxi Li, Chunhua Shen, and Qinfeng Shi, Real-time visual tracking using compressive sensing, in *IEEE Conference on Computer Vision and Pattern Recognition (CVPR)*, pp. 1305–1312, 2011. DOI: 10.1109/cvpr.2011.5995483 42

[102] Kaihua Zhang, Lei Zhang, and Ming-Hsuan Yang, Real-time compressive tracking, in *European Conference on Computer Vision*, pp. 864–877, Springer, 2012. DOI: 10.1007/978-3-642-33712-3_62 42, 57

[103] Paul Viola and Michael Jones, Rapid object detection using a boosted cascade of simple features, in *Proc. of the IEEE Computer Society Conference on Computer Vision and Pattern Recognition (CVPR)*, vol. 1, p. I, 2001. DOI: 10.1109/cvpr.2001.990517 42

[104] Rudolph Kalman, A new approach to linear filter and prediction theory 3, in *Basic. Engr. D*, vol. 82, pp. 35–45, 1960. DOI: 10.1115/1.3662552 43

[105] Jayaraman J. Thiagarajan, Karthikeyan N. Ramamurthy, Peter Knee, Andreas Spanias, and Visar Berisha, Sparse representations for automatic target classification in SAR images, in *4th IEEE International Symposium on Communications, Control and Signal Processing (ISCCSP)*, Limassol, Cyprus, 2010. DOI: 10.1109/isccsp.2010.5463416 44

[106] Maria Valera and Sergio A. Velastin, Intelligent distributed surveillance systems: A review, in *IEE Proceedings-Vision, Image and Signal Processing*, vol. 152, no. 2, pp. 192–204, 2005. DOI: 10.1049/ip-vis:20041147 44

[107] Sameeksha Katoch, Gowtham Muniraju, Sunil Rao, Andreas Spanias, Pavan Turaga, Cihan Tepedelenlioglu, Mahesh Banavar, and Devarajan Srinivasan, Shading prediction, fault detection, and consensus estimation for solar array control, in *IEEE Industrial Cyber-Physical Systems (ICPS)*, pp. 217–222, St. Petersburg, Russia, 2018. DOI: 10.1109/icphys.2018.8387662 44

[108] Jing Liu, Feng Lian, and Mahendra Mallick, Distributed compressed sensing based joint detection and tracking for multistatic radar system, in *Information Sciences*, vol. 369, pp. 100–118, 2016. DOI: 10.1016/j.ins.2016.06.032 44

[109] Kuldeep Kulkarni and Pavan Turaga, Reconstruction-free action inference from compressive imagers, in *IEEE Transactions on Pattern Analysis and Machine Intelligence*, vol. 38, no. 4, pp. 772–784, 2016. DOI: 10.1109/tpami.2015.2469288 46, 48

[110] Henry Braun, Image Reconstruction, Classification, and Tracking for Compressed Sensing Imaging and Video, Ph.D. thesis, Arizona State University, November, 2016. 48

[111] D. J. Salmond and H. Birch, A particle filter for track-before-detect, in *Proc. of IEEE American Control Conference*, vol. 5, pp. 3755–3760, 2001. DOI: 10.1109/acc.2001.946220 51

[112] Vision@reading. PETS: Performance evaluation of tracking and surveillance, 2007. http://www.cvg.rdg.ac.uk/slides/pets.html 52

[113] Nil Goyette, Pierre-Marc Jodoin, Fatih Porikli, Janusz Konrad, and Prakash Ishwar, Changedetection.net: A new change detection benchmark dataset, in *IEEE Computer Society Conference on Computer Vision and Pattern Recognition Workshops*, pp. 1–8, 2012. DOI: 10.1109/cvprw.2012.6238919 52, 54

[114] Sam Hare, Amir Saffari, and Philip H. S. Torr, Struck: Structured output tracking with Kernels, in *IEEE International Conference on Computer Vision (ICCV)*, pp. 263–270, 2011. DOI: 10.1109/iccv.2011.6126251 57

[115] Aswin C. Sankaranarayanan, Pavan K. Turaga, Richard G. Baraniuk, and Rama Chellappa, Compressive acquisition of dynamic scenes, in *ECCV*, pp. 129–142, Springer, 2010. DOI: 10.1007/978-3-642-15549-9_10 59

[116] K. Kulkarni and P. Turaga, Reconstruction-free action inference from compressive imagers, in *Pattern Analysis and Machine Intelligence, IEEE Transactions on*, vol. PP, no. 99, 2015. DOI: 10.1109/tpami.2015.2469288 59

[117] Suhas Lohit, Kuldeep Kulkarni, Pavan Turaga, Jian Wang, and Aswin C. Sankaranarayanan, Reconstruction-free inference on compressive measurements, in *4th International Conference on Computational Cameras and Displays, Held in Conjunction with IEEE CVPR*, June 2015. DOI: 10.1109/cvprw.2015.7301371 59

[118] Yun Li, Chinmay Hegde, Aswin C. Sankaranarayanan, Richard Baraniuk, and Kevin F. Kelly, Compressive image acquisition and classification via secant projections, in *Journal of Optics*, vol. 17, no. 6, pp. 065701, 2015. DOI: 10.1088/2040-8978/17/6/065701 60

[119] Suhas Lohit, Kuldeep Kulkarni, and Pavan Turaga, Direct inference on compressive measurements using convolutional neural networks, in *IEEE International Conference on Image Processing (ICIP)*, pp. 1913–1917, 2016. DOI: 10.1109/icip.2016.7532691 60

[120] Yann LeCun, Léon Bottou, Yoshua Bengio, and Patrick Haffner, Gradient-based learning applied to document recognition, in *Proc. of the IEEE*, vol. 86, no. 11, pp. 2278–2324, 1998. DOI: 10.1109/5.726791 60

[121] Olga Russakovsky, Jia Deng, Hao Su, Jonathan Krause, Sanjeev Satheesh, Sean Ma, Zhiheng Huang, Andrej Karpathy, Aditya Khosla, Michael Bernstein, and A.C. Berg, Imagenet large scale visual recognition challenge, in *International Journal of Computer Vision*, vol. 115, no. 3, pp. 211–252, 2015. DOI: 10.1007/s11263-015-0816-y 60

[122] Henry Braun, Pavan Turaga, Andreas Spanias, and Cihan Tepedelenlioglu, Direct classification from compressively sensed images via deep Boltzmann machine, in *50th IEEE Asilomar Conference on Signals, Systems and Computers*, pp. 454–457, Pacific Grove, CA, November 2016. DOI: 10.1109/acssc.2016.7869080 61

[123] Stanislaw Antol, Aishwarya Agrawal, Jiasen Lu, Margaret Mitchell, Dhruv Batra, C Lawrence Zitnick, and Devi Parikh, VQA: Visual question answering, in *Proc. of the IEEE International Conference on Computer Vision*, pp. 2425–2433, 2015. DOI: 10.1109/iccv.2015.279 61

[124] Christian Szegedy, Wei Liu, Yangqing Jia, Pierre Sermanet, Scott Reed, Dragomir Anguelov, Dumitru Erhan, Vincent Vanhoucke, and Andrew Rabinovich, Going deeper with convolutions, in *Proc. of the IEEE Conference on Computer Vision and Pattern Recognition*, pp. 1–9, 2015. DOI: 10.1109/cvpr.2015.7298594 62

[125] Tomas Mikolov, Ilya Sutskever, Kai Chen, Greg S. Corrado, and Jeff Dean, Distributed representations of words and phrases and their compositionality, in *Advances in Neural Information Processing Systems*, pp. 3111–3119, 2013. 62

[126] Li-Chi Huang, Kuldeep Kulkarni, Anik Jha, Suhas Lohit, Suren Jayasuriya, and Pavan Turaga, CS-VQA: Visual question answering with compressively sensed images, in *25th IEEE International Conference on Image Processing (ICIP)*, pp. 1283–1287, 2018. DOI: 10.1109/icip.2018.8451445 61, 62, 63

[127] Yash Goyal and Tejas Khot and Douglas Summers-Stay and Dhruv Batra and Devi Parikh, Making the V in VQA Matter: Elevating the Role of Image Understanding in Visual Question Answering, in *Conference on Computer Vision and Pattern Recognition (CVPR)*, 2017. 63

[128] Aurora Schmidt, Joel B. Harley, and José M. F. Moura, Compressed sensing radar surveillance networks, in *7th IEEE Sensor Array and Multichannel Signal Processing Workshop (SAM)*, pp. 209–212, 2012. DOI: 10.1109/sam.2012.6250469 66

[129] Darren Craven, Brian McGinley, Liam Kilmartin, Martin Glavin, and Edward Jones, Compressed sensing for bioelectric signals: A review, in *IEEE Journal of Biomedical and Health Informatics*, vol. 19, no. 2, pp. 529–540, 2015. DOI: 10.1109/jbhi.2014.2327194 66

[130] Christopher A. Metzler, Arian Maleki, and Richard G. Baraniuk, From denoising to compressed sensing, in *IEEE Transactions on Information Theory*, vol. 62, no. 9, pp. 5117–5144, 2016. DOI: 10.1109/tit.2016.2556683

[131] Joachim H. G. Ender, On compressive sensing applied to radar, in *Signal Processing*, vol. 90, no. 5, pp. 1402–1414, 2010. DOI: 10.1016/j.sigpro.2009.11.009

[132] Saad Qaisar, Rana Muhammad Bilal, Wafa Iqbal, Muqaddas Naureen, and Sungyoung Lee, Compressive sensing: From theory to applications, a survey, in *Journal of Communications and Networks*, vol. 15, no. 5, pp. 443–456, 2013. DOI: 10.1109/jcn.2013.000083

[133] Karthikeyan Natesan Ramamurthy and Andreas Spanias, Optimized measurements for Kernel compressive sensing, in *45th IEEE Asilomar Conference on Signals, Systems and Computers*, pp. 1443–1446, Pacific Grove, CA, November 2011. DOI: 10.1109/acssc.2011.6190256 66

[134] Richard J. Radke, Srinivas Andra, Omar Al-Kofahi, and Badrinath Roysam, Image change detection algorithms: A systematic survey, in *IEEE Transactions on Image Processing*, vol. 14, no. 3, pp. 294–307, 2005. DOI: 10.1109/tip.2004.838698 66

[135] Florian Barbieri, Sumedha Rajakaruna, and Arindam Ghosh, Very short-term photovoltaic power forecasting with cloud modeling: A review, in *Renewable and Sustainable Energy Reviews*, vol. 75, pp. 242–263, 2017. DOI: 10.1016/j.rser.2016.10.068

[136] Jayaraman J. Thiagarajan, Karthikeyan Natesan Ramamurthy, Pavan Turaga, and Andreas Spanias, *Image Understanding Using Sparse Representations*, vol. 7, Morgan & Claypool Publishers, Ed., Al Bovic, 2014.

[137] H. Braun, S. T. Buddha, V. Krishnan, C. Tepedelenlioglu, A. Spanias, M. Banavar, and D. Srinivasan, Topology reconfiguration for optimization of photovoltaic array output, in *Sustainable Energy, Grids and Networks*, vol. 6, pp. 58–69, 2016. DOI: 10.1016/j.segan.2016.01.003

[138] Karthikeyan Natesan Ramamurthy, Jayaraman J. Thiagarajan, and Andreas Spanias, Improved sparse coding using manifold projections, in *18th IEEE International Conference on Image Processing*, pp. 1237–1240, Belgium, 2011. DOI: 10.1109/icip.2011.6115656

[139] Henry Braun, Santoshi Tejasri Buddha, Venkatachalam Krishnan, Cihan Tepedelenlioglu, Andreas Spanias, Toru Takehara, Ted Yeider, Mahesh Banavar, and Shinichi Takada, Signal processing for solar array monitoring, fault detection, and optimization, in *Synthesis Lectures on Power Electronics*, vol. 7, no. 1, pp. 1–95, 2012. DOI: 10.2200/s00425ed1v01y201206pel004 66

[140] Sunil Rao, Sameeksha Katoch, P. Turaga, Andreas Spanias, Cihan Tepedelenlioglu, Raja Ayyanar, Henry Braun, Jongmin Lee, U. Shanthamallu, M. Banavar, and D. Srinivasan, A cyber-physical system approach for photovoltaic array monitoring and control, in *8th IEEE International Conference on Information, Intelligence, Systems and Applications (IISA)*, pp. 1–6, Larnaca, August 2017. DOI: 10.1109/iisa.2017.8316458 66

[141] Rohit Ranjan Verma Mohit, Sameeksha Katoch, Ashoka Vanjare, and S. N. Omkar, Classification of complex UCI datasets using machine learning algorithms using hadoop, in *Internation Journal of Computer Science and Software Engineering (IJCSSE)*, vol. 4, no. 7, 2015.

[142] Sameeksha Katoch, Pavan Turaga, Andreas Spanias, and Cihan Tepedelenlioglu, Fast non-linear methods for dynamic texture prediction, in *25th IEEE International Conference on Image Processing (ICIP)*, pp. 2107–2111, Athens, Greece, 2018. DOI: 10.1109/icip.2018.8451479

[143] Huan Song, Deepta Rajan, Jayaraman J. Thiagarajan, and Andreas Spanias, Attend and diagnose: Clinical time series analysis using attention models, in *32nd AAAI Conference on Artificial Intelligence*, New Orleans, Louisiana, LA, 2018. 66

[144] Jürgen Schmidhuber, Deep learning in neural networks: An overview, in *Neural Networks*, vol. 61, pp. 85–117, 2015. DOI: 10.1016/j.neunet.2014.09.003 66

[145] Yann LeCun, Yoshua Bengio, and Geoffrey Hinton, Deep learning, in *Nature*, vol. 521, no. 7553, p. 436, 2015. DOI: 10.1038/nature14539

[146] Martin Längkvist, Lars Karlsson, and Amy Loutfi, A review of unsupervised feature learning and deep learning for time-series modeling, in *Pattern Recognition Letters*, vol. 42, pp. 11–24, 2014. DOI: 10.1016/j.patrec.2014.01.008

[147] Uday Shankar Shanthamallu, Andreas Spanias, Cihan Tepedelenlioglu, and Mike Stanley, A brief survey of machine learning methods and their sensor and IoT applications, in *8th IEEE International Conference on Information, Intelligence, Systems and Applications (IISA)*, pp. 1–8, Larnaca, August 2017. DOI: 10.1109/iisa.2017.8316459 66

Authors' Biographies

HENRY BRAUN

Henry Braun is a researcher at Magnetic Resonance Research center at University of Minnesota. He received his Ph.D. in electrical engineering from Arizona State University in 2016. His research interests include computer vision, signal processing, and compressive sensing.

PAVAN TURAGA

Pavan Turaga (S'05, M'09, SM'14) is an associate professor in the School of Arts, Media, Engineering, and Electrical Engineering at Arizona State University. He received a B.Tech. degree in electronics and communication engineering from the Indian Institute of Technology Guwahati, India, in 2004, an the M.S. and Ph.D. in electrical engineering from the University of Maryland, College Park in 2008 and 2009, respectively. He then spent two years as a research associate at the Center for Automation Research, University of Maryland, College Park. His research interests are in imaging and sensor analytics with a theoretical focus on non-Euclidean and high-dimensional geometric and statistical techniques. He was awarded the Distinguished Dissertation Fellowship in 2009. He was selected to participate in the Emerging Leaders in Multimedia Workshop by IBM, New York, in 2008. He received the National Science Foundation CAREER award in 2015.

ANDREAS SPANIAS

Andreas Spanias is Professor in the School of Electrical, Computer, and Energy Engineering at Arizona State University. He is also the director of the Sensor Signal and Information Processing (SenSIP) center and the founder of the SenSIP industry consortium (also an NSF I/UCRC site). His research interests are in the areas of adaptive signal processing, speech processing, machine learning, and sensor systems. He and his student team developed the computer simulation software Java-DSP and its award-winning iPhone/iPad and Android versions. He is author of two textbooks: *Audio Processing and Coding* by Wiley and *DSP: An Interactive Approach* (2nd Ed.). He contributed to more than 300 papers, 7 monographs, 9 full patents, 6 provisional patents, and 10 patent pre-disclosures. He served as Associate Editor of the *IEEE Transactions on Signal Processing* and as General Co-chair of *IEEE ICASSP-99*. He also served as the IEEE Signal Processing Vice-President for Conferences. Andreas Spanias is co-recipient of the 2002 IEEE Donald G. Fink paper prize award and was elected Fellow of the IEEE in 2003. He served as Distinguished Lecturer for the IEEE Signal processing society in 2004. He is a series editor for the Morgan and Claypool lecture series on algorithms and software. He received recently the 2018 IEEE Phoenix Chapter award with citation: "For significant innovations and patents in signal processing for sensor systems." He also received the 2018 IEEE Region 6 Educator Award (across 12 states) with citation: "For outstanding research and education contributions in signal processing."

SAMEEKSHA KATOCH

Sameeksha Katoch is a Ph.D.student in the School of Electrical, Computer, and Energy Engineering at Arizona State University. She received a B.Tech. Degree in electronics and communication engineering from the National Institute of Technology Srinagar, India, in 2015, and a M.S. degree in electrical engineering from the Arizona State University in 2018. She has received IEEE Al Gross Student Award. Her research interests are in computer vision, signal processing, and deep learning.

SUREN JAYASURIYA

Suren Jayasuriya has been an assistant professor at Arizona State University since 2018. He was a postdoctoral fellow at the Robotics Institute at Carnegie Mellon University, U.S. in 2017. He received a B.S. in mathematics and a B.A. in philosophy from the University of Pittsburgh, in 2012, and received his Ph.D. in Electrical and Computer Engineering from Cornell University, in 2017. His research interests are in computational photography and imaging, computer vision, and image sensors.

CIHAN TEPEDELENLIOGLU

Cihan Tepedelenlioglu (S'97-M'01) was born in Ankara, Turkey in 1973. He received his B.S. with highest honors from Florida Institute of Technology in 1995, and his M.S. from the University of Virginia in 1998, both in Electrical Engineering. From January 1999 to May 2001 he was a research assistant at the University of Minnesota, where he completed his Ph.D. in Electrical and Computer Engineering. He is currently an Associate Professor of Electrical Engineering at Arizona State University. He was awarded the NSF (early) Career grant in 2001, and has served as an Associate Editor for several IEEE Transactions including *IEEE Transactions on Communications,* *IEEE Signal Processing Letters,* and *IEEE Transactions on Vehicular Technology.* His research interests include statistical signal processing, system identification, wireless communications, estimation and equalization algorithms for wireless systems, multi-antenna communications, OFDM, ultra-wideband systems, distributed detection and estimation, and data mining for PV systems.

Printed in the United States
by Baker & Taylor Publisher Services